GETTING STARTED
WITH THE TI-92/92 PLUS
GRAPHING CALCULATOR

GETTING STARTED WITH THE TI-92/92 PLUS GRAPHING CALCULATOR

Carl Swenson
Seattle University

Brian Hopkins
Seattle University

John Wiley & Sons, Inc.
New York • Chichester • Weinheim • Brisbane • Singapore • Toronto

Cover image by Marjory Dressler

Copyright © 1998 by John Wiley & Sons, Inc.

All rights reserved.

Reproduction or translation of any part of this work beyond that permitted by Sections 107 and 108 of the 1976 United States Copyright Act without the permission of the copyright owner is unlawful. Requests for permission or further information should be addressed to the Permissions Department, John Wiley & Sons, Inc.

ISBN 0-471-25364-2

Printed in the United States of America

10 9 8 7 6 5 4 3 2 1

Printed and bound by Hamilton Printing Company
Cover printed by The Lehigh Press, Inc.

PREFACE

The purpose of this book is to show how to apply the features of the TI-92 and the TI-92 Plus graphing calculators to understand calculus. If at all possible, use a TI-92 Plus, particularly if you plan to do the differential equation part. The TI-89 had been announced but was not available for testing at the time of publication. We have been informed that the TI-89 and the TI-92 Plus are nearly equivalent, except for the screen and keyboard size. So although the name TI-89 is not in the title, this book can be used for that calculator as well.

The book is divided into five parts, corresponding to common areas of focus in a calculus course. The chapters provide a more specific description of each calculus topic. In general, if you are looking for help on a calculus topic, then use the Table of Contents to find the topic, but if you are looking for help on a calculator command, then look in the Index. Each calculus chapter is intended to stand alone, but they all require an understanding of the basics from Part I Precalculus. This is intended to be a review; it can be skimmed by experienced users or used as a primer by those new to this calculator.

Demonstrations are presented at the end of each part. These show off the special features of the calculator; a few are more motivated by the calculator than the mathematics.

Most of the examples are taken from the widely used calculus book, *Calculus* by Deborah Hughes-Hallett, Gleason, et al. We would like to acknowledge and thank the Calculus Consortium based at Harvard and John Wiley & Sons, Inc. for permission to freely use examples from that text.

To the student

Using a graphing calculator can be both frustrating and fun. A healthy attitude when you get frustrated is to step back and say, "Isn't that interesting that it doesn't work." Figuring out how things work can be fun. If you get too frustrated, then it is time to ask a friend or the instructor for help. Make sure you have a phone list of friends with the same calculator.

Part I gives you clear sets of key sequences so that you will become comfortable with how your calculator works. The remaining parts shift into a higher gear and only show you calculator screens as guides for the keystrokes. Your TI-92 *Guidebook* provides a resource if you get stuck; it explains each feature briefly, usually with a key sequence example.

Remember that the *Guidebook* is like a dictionary: there is no story line or context. In this book, the features that you need for calculus will be explained in the context of calculus examples. Other calculator features that are less important to calculus may not be mentioned at all. The mathematical content drives this presentation, not the calculator features (with the occasional exception in the demonstrations).

This calculator has a symbolic algebra system and many advanced features not available on earlier TI graphing calculators. In order to minimize the number of steps and possible errors in our examples, we use the default modes and setting whenever possible. The functions and examples are quite simplistic; they are meant to get you started. Once you have a feature working for a known case, you can branch out more safely into greater complexity.

We have included tips about such things as short-cuts, warnings, and related ideas. We hope you will find them useful.

➤ *Tip: Don't use technology in place of thinking.*

To the instructor

It has been our aim to write a set of parallel form books that span the TI graphing calculator series so that you can allow a variety of these models in your classes. These materials are designed to allow you to focus on the calculus, not the calculator. By having the students use these calculator-specific materials, you should be able to greatly reduce the problems of using multiple TI models.

Will these materials take care of all your students? Of course not. There will still be the zealous ones who want the programs in assembly language and the anxious ones who want the buttons pressed for them. These materials are aimed at the middle, giving enough guidance so that most students will be able to work through an example without assistance, but not so specific as to be considered a mindless exercise in pressing keys in the right order.

Programming is not an emphasis of this text. We have included only one serious program which offers graphical help in understanding the Riemann sum. We suggest that you acquire the **TI-GRAPH LINK**™ and download this program from the TI archives on the Internet where it is available. (See the Appendix for the internet address.) You can distribute programs and data to your class. Should you be simultaneously using other versions of this book (for the TI-86 and below), you will see that those have four programs. The power of the TI-92 has eliminated the need for programs to make up for missing features, such as Taylor polynomial generation. The Riemann sum program has been written with cross-platform compatibility in mind, and an effort was made so that the programs are almost identical throughout the book series.

➤ *Tip: The TI Volume Purchase Plan can provide you with the* **TI-GRAPH LINK** ™ *package and/or an overhead model for classroom use.*

Thanks

The first author would like to thank Seattle University for its generous sabbatical support during the 1997-1998 academic year which allowed me to prepare these materials. Also I would like to thank my son Will who provided his computer for editing. My greatest thanks go to my wife Julia Buchholz for her patience and constant support.

The second author would like to thank the friends and students that help enrich his life, especially David.

We would both like to express our appreciation for the help and support of our colleagues in the Seattle University Mathematics Department. They have all shared with us their experiences of using the TI calculators in various classes. They are Mary Ehlers, Wynne Guy, Brian Henderson, Janet Mills (Chair), Ahmad Mirbagheri, Kathy Sullivan, Donna Sylvester, Alan Troy and André Yandl. A special thanks to Ahmad for his suggestions of activities and the use of his materials.

The Seattle University Information Services Help Desk staff was patient and helpful with technical computer problems.

Texas Instruments was helpful in loaning us TI-92 Plus chips in advance of production. Their staff answered our questions and gave us the support we needed. Special thanks to Michelle Miller.

The folks at Wiley have been a great help in answering questions and making arrangements. Special thanks to Ruth Baruth who suggested this project; to Sharon Smith, the editor; and to Mary Johenk and Madelyn Lesure for handling details.

Carl Swenson, *swenson@seattleu.edu*

Brian Hopkins, *hopkins@seattleu.edu*

CONTENTS

PART I PRECALCULUS

1. BASIC CALCULATIONS   2

   Getting started with basic keys 2
   The HOME screen 3
   Entering expressions and instructions 4
   Magic tricks to change the keyboard: 2nd, ♦, and ↑ 5
   Getting around: navigation and editing keys 6
   The format of numbers 8
   How to put yourself in a good MODE 9
   Stored values and symbolic variables 11

2. FUNCTIONS: A FUNDAMENTAL TOOL FOR CALCULUS   14

   Formula vs. function notation: using the Y= editor 14
   A summary of menu use 16
   Evaluating a function at a point 17
   New functions from old 17
   Defining families of functions by using lists: y1={...}x 18
   A note about defining functions on a TI-92 without Plus 19

3. TABLES OF FUNCTION VALUES   20

   Lists of function values: f({...}) 20
   A table of values for a function: TABLE 20
   A table for selected functions 21
   Find the zero of a function from a table 22
   Editing a function formula from inside a table 23

4. GRAPHING A FUNCTION   24

Basic graphing: ♦_Y=, ♦_WINDOW and ♦_GRAPH 24

Finding a good window: Zoom 25

Identifying points on the screen 27

Reading function values from the graph: Trace 27

Panning a window 29

What is a good window? 31

Can you always find a good window? 31

Other Zoom options 34

5. EXTENDED GRAPHING FEATURES   35

Finding roots: Zero 35

Finding extrema: Minimum and Maximum 36

Finding an intersection of two graphs 37

Split screen graphing (optional) 38

Graphing inverse functions 39

Putting text on graphs 40

6. SOLVING EQUATIONS   41

Solving single variable equations 41

What if ? analysis of an investment using Numerical Solver 46

DEMONSTRATIONS   48

A. THE SINE ANIMATION 48

B. THE TEXT EDITOR 50

C. COBB-DOUGLAS 3D GRAPHING 52

PART II   DIFFERENTIAL CALCULUS

7. THE LIMIT CONCEPT   56

Creating data lists and finding average velocity 56

What does the limit mean graphically and numerically? 57

Speeding ticket: the Math Police let you off with a warning 59

The `Limit` function 59
One-sided limits (optional) 61

8. FINDING THE DERIVATIVE AT A POINT — 62

Comparing derivatives at a point: three methods 62
Comparing the exact and the numeric derivative 65

9. THE DERIVATIVE AS A FUNCTION — 68

Comparing the numeric to the symbolic derivative 68
Viewing a graph of derivative function 69
The function that is its own derivative: $y = e^x$ 71
The symbolic derivative of common functions 72

10. THE SECOND DERIVATIVE: THE DERIVATIVE OF THE DERIVATIVE — 73

Symbolic derivatives and their graphs 73
Looking at the concavity of the logistic curve 74
Creating a numeric second derivative using a table 76

11. THE RULES OF DIFFERENTIATION — 78

The Product Rule 78
The Quotient Rule 79
The Chain Rule 81
The derivative of the tangent function 81

12. OPTIMIZATION — 83

The ladder problem 83
Box with lid 87
Using the second derivative to find concavity 88

DEMONSTRATIONS — 90

A. OPTIMIZATION WITH CABRI GEOMETRY 90
B. IMPLICIT DIFFERENTIATION 94

PART III   INTEGRAL CALCULUS

13. LEFT- AND RIGHT-HAND SUMS — 96

Distance from the sum of the velocity data   96
Summing lists to create left- and right-hand sums: Σ 97
Approximating area using the left- and right-hand sums   99

14. THE DEFINITE INTEGRAL — 100

The definite integral from a graph: $\int f(x)$ 100
The definite integral as a number on the **HOME** screen   101
Facts about the definite integral   101
The definite integral as a function: $y=\int(\ldots,x)$ 103

15. THE INDEFINITE INTEGRAL AND THE FUNDAMENTAL THEOREM — 105

The antiderivative   105
The Fundamental Theorem   105
Comparing $d(\int(\ldots)\ldots)$ and $\int(d(\ldots)\ldots)$ 108

16. RIEMANN SUMS — 110

A few words about programs   110
Using the `rsum()` program to find Riemann sums   112
The TI-92 program `rsum()`   114

17. IMPROPER INTEGRALS — 117

An infinite limit of integration   117
The integrand goes infinite   119
The comparison test   120

18. APPLICATIONS OF THE INTEGRAL — 121

Geometry: arc length   121
Physics: force and pressure   122

Economics: present and future value   123
Modeling: normal distributions   125

DEMONSTRATIONS   128

A. INTEGRATION BY PARTIAL FRACTIONS   128
B. THE USE AND CONVERSION OF UNITS   130

PART IV   SERIES

19. TAYLOR SERIES AND SERIES CONVERGENCE   136

The Taylor polynomials for $y = e^x$ 136
The Taylor series for $\ln(x)$ 137
The Taylor series of the sine function   138
How can we know if a series converges?   139

20. GEOMETRIC SERIES   142

The general formula for a finite geometric series   142
Identifying the parameters of an infinite geometric series   144
Summing an infinite series by the formula   145
Piggy-bank vs. trust   146

21. FOURIER SERIES   148

A word about user-defined functions   148
User-defined functions   150
The general formula for the Fourier approximation function   151
A system for graphing the Fourier approximation function   152

DEMONSTRATION

A. INFINITE SERIES   155

PART V   DIFFERENTIAL EQUATIONS

22. DIFFERENTIAL EQUATIONS AND SLOPE FIELDS   158

A note on the necessity of the TI-92 Plus   158
The solutions of differential equations   158
The discrete learning curve: a sequence function   159
The continuous learning curve $y' = 100 - y$ 160
The analytic solution of a differential equation   162
Slope fields for differential equations   163

23. EULER'S METHOD   166

The relationship of a differential equation to a difference equation   166
Euler's method for $y' = -x/y$ starting at (0,1)   167
Euler gets lost going around a corner   169

24. SECOND-ORDER DIFFERENTIAL EQUATIONS   170

The second-order equation $s'' = -g$ 170
The second-order equation $s'' + \omega^2 s = 0$ 172
The linear second-order equation $y'' + by' + cy = 0$ 173

25. THE LOGISTIC POPULATION MODEL   177

Using the `Data Editor` for entering US population data   177
Using `Plots` to graph data lists   178
Fitting data with a logistic equation   179
The logistic curve   180
Using the regression line to rewrite the differential equation   182

26. SYSTEMS OF EQUATIONS AND THE PHASE PLANE   184

The S-I-R model   184
Predator-prey model   187

DEMONSTRATION 190

    A. EULER'S METHOD ON A SYSTEM OF EQUATIONS   190

APPENDIX AND INDEX

APPENDIX 194

    Complex number form   194
    Polar coordinates in the complex plane   194
    Parametric graphing   196
    Internet address information   197
    Linking calculators   198
    Linking to a computer   198
    Troubleshooting   199

INDEX 203

PART I
PRECALCULUS

1. BASIC CALCULATIONS
2. FUNCTIONS: A FUNDAMENTAL TOOL FOR CALCULUS
3. TABLES OF FUNCTION VALUES
4. GRAPHING A FUNCTION
5. EXTENDED GRAPHING FEATURES
6. SOLVING EQUATIONS

DEMONSTRATIONS:
- A. THE SINE ANIMATION
- B. THE TEXT EDITOR
- C. COBB-DOUGLAS 3D GRAPHING

1. BASIC CALCULATIONS

The tools used to make numeric calculations have developed from the fingers, to an abacus, to a slide rule, to a scientific calculator, and now to a calculator capable of symbolic manipulation. In this chapter we see how to use the TI-92 — a current tool of calculation. If you have used a graphing calculator before, you may only need to skim this chapter to understand the differences between your old and new calculator. Chapter 2 of the TI-92 *Guidebook* should also be consulted if you are having difficulty getting started. This chapter will be a brief summary of that material. In this book you will find the references to TI calculator keys and menu choices written in the TI-92 font. `The TI-92 font looks like this.`

A note about the TI-92 Plus and the TI-89

If you are a TI-92 user and intend to use the calculator for differential equations, then it will be worthwhile to upgrade the TI-92 to a TI-92 Plus. (This requires a simple addition of a chip to your existing calculator.) Screen illustrations in this book are from a TI-92 Plus. The screens of the TI-92 without the Plus are similar, but not always exactly the same. The TI-92 Plus has more features, so its menus are sometimes longer. When there is a major difference, it will be pointed out in a parenthetic note or in a separate section such as this.

The TI-89 has been announced and is intended to be functionally equivalent to the TI-92 Plus, except that it will not have Cabri geometry capabilities. Although the screen size and keyboard are different, the basic computations and commands in this book should be the same on the TI-89.

Getting started with basic keys

This calculator was designed to be held with both hands. Place your left thumb in the indentation at the upper left corner and your right thumb in the center of the blue navigation key. The calculator can also be laid flat or propped up with the indentations in the cover to free your hands. The keyboard basically has three areas: on the right hand side are the calculation and navigation keys, below the screen is a QWERTY keyboard and in the upper left is a set of blue function keys to access pull-down menus.

The `ON` key

Study the keyboard and press the `ON` key in the lower left hand corner. You will probably see a menu bar across the top and a blinking vertical line cursor in the lower left of the screen. If you do not, then you may need to set the screen contrast. Even if your screen is showing, it is a good idea to know how to set the screen contrast: as you use the calculator, the battery will wear down and it will be necessary to adjust the screen. Also, the screen contrast may need to be adjusted for different lighting environments.

Adjusting the screen contrast

To darken the screen, press and hold the green diamond key (♦) next to ON and then repeatedly press the black plus key on the right of the keypad. The screen can be lightened by but using the black subtraction key instead of the plus key. If the setting is too low, the screen will not show, and if it is too high, the screen will be dark as night.

If you take a break and come back, the screen will have disappeared for a different reason. The calculator "goes to sleep" and turns itself off after a few minutes of no activity. Although there is an OFF key, there is little reason to use it.

➤ *Tip: Sometimes "broken" calculators can be fixed by reinserting the batteries correctly.*

The HOME screen

On most graphing calculators, results flow down the screen as you work. The TI-92 works differently. Commands are entered from a lower entry line and answers scroll up the screen into the history area. Keep your eye on the highlighting or blinking cursor to know where you are. The four areas on the HOME screen from top to bottom are

- *Toolbar* — Top row: pull down menus activated by the corresponding blue function keys, F1 to F8.
- *History* — Main screen: displays previous instructions and answers. Use up-arrow to scroll and highlight instructions or answers. They can be copied to the entry line.
- *Entry line* — Action line: input and edit expressions or instructions.
- *Status line* — Bottom row: current settings and messages in fine print.

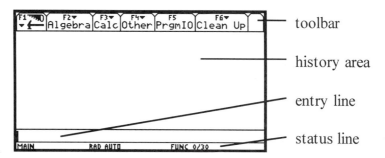

Figure 1.1 The HOME screen areas.

Entering expressions and instructions

On the right side of the calculator are the essential keys used on a scientific calculator. Master these first. The gray numeric keys are used for simple arithmetic calculations. There are three **ENTER** keys for your convenience; they all do the same thing. In this chapter, keystrokes will be given for the results shown on the screens.

➤ *Tip: Test technology with known results before trying complex examples.*

We start by typing

3 ÷ 2 ENTER

Be aware that the symbol on the divide key, ÷, is different from the divide symbol (/) on the screen. The screen shows the input on the left and output on the right. Notice these are the same in this case because, unlike most calculators, the TI-92 responds with the exact form instead of giving a decimal answer. (We will shortly learn how to put answers in decimal format.)

➤ *Tip: Some keypad symbols, like ÷, are displayed on the screen with a different symbol from that on the key.*

Make special note that the gray (-) key is for negation and must be distinguished from the black − subtraction key. One of the most common errors is interchanging the use of the subtraction key and the negation key. Give it a moment's thought: subtraction requires two numbers, while negation works on a single number. Try pressing the following sequences of four keys.

3 − 2 ENTER

and then

3 (−) 2 ENTER

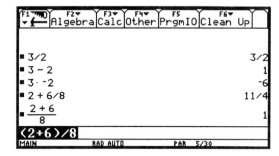

Figure 1.2 Division symbol, and subtraction vs. negation key distinction.

Look carefully at Figure 1.2 which shows these two symbols: subtraction is longer and centered, while negation is shorter and raised. The first expression gave 1 and the second entry gave -6 as an answer. An important feature of this calculator is that it formats both input and answers (this feature is called *pretty print*). By looking at the screen you can see that it has placed a multiplication dot between the two numbers to show you that it multiplied 3 times -2 when you entered 3(−)2.

➤ *Tip: Before using an answer, check the input format in the history area to see that you entered the expression correctly. The expression is still highlighted on the entry line and can be easily edited and recalculated if necessary.*

The single most popular error (can errors be popular?) among new users is to not use parentheses when needed. This is a serious error because the calculator does not stop and alert you with an error screen; instead, it gives you the correct answer to a question you are not asking. Suppose you want to add 2 and 6 and then divide by 8. We don't need a calculator to tell us the expression has value 1. But if you enter 2 + 6 / 8, the answer will be 11/4. You can figure out that the calculator divided 6 by 8 first and then added that to 2. This was not what we wanted. We need to use parentheses to insure that we are evaluating the correct expression. Now try (2 + 6) / 8 and get an answer of 1 as expected. Note how the expression is shown in the history area as a fraction. There is a prescribed order of operations on your calculator and the pretty print feature shows an unambiguous expression and its answer (if you want more details, see page 492 of the *Guidebook*).

➤ *Tip: When you get an unexpected result, go back and check parentheses. Be generous; adding extra parentheses doesn't hurt.*

Magic tricks to change the keyboard: 2nd, ♦, and ↑

How can we do more with the basic keys? The trick is modifier keys, like shift and control on a computer keyboard. A difference from a computer keyboard is that the modifier keys on the calculator are pressed once and not held down. The indicator of a modified condition is on the status line just to the right of the word MAIN. The first modifier key is the yellow 2nd key. After pressing it once (presto!) all the keys now have a new meaning. These meanings are indicated in yellow just above each key. For a function such as SIN or LN, the 2nd key gives its inverse. The notation 2nd_e^x is used in this book to denote that e^x is an entry that needs to be first modified by the 2nd key.

We have already used the green ♦ key to adjust the screen contrast. It also provides easy access to graph and table commands with the Q, W, E, R, T, Y keys. (This provides familiarity for users with experience using older TI calculator.) For the TI-92 Plus, pressing ♦ and then the up arrow moves the cursor to the beginning of the entry line; using the down arrow instead moves the cursor to the end of the line.

The third modifier key is the white ↑ key. It is used like a shift key on a computer to enter uppercase letters.

➤ *Tip: The* **2nd**, ♦, *and* ↑ *keys work as toggles: if you press one by mistake and turn on a modifier, just press the key again to turn it off.*

The greatest equation ever written: $e^{i\pi} + 1 = 0$

Exponentiation is written in mathematics as a superscript; this is easy by hand, but not directly possible on a calculator. The caret ^ is used to signify exponents. Let's practice. We know $2^3 = 8$ and to verify this on our calculator, we type

$$2 \; \wedge \; 3 \; \text{ENTER}$$

You will see this expression reformatted into mathematical exponentiation style. Look in the history area in Figure 1.3 to see how this exponent appears.

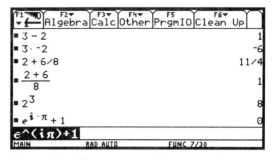

Figure 1.3 *The equation* $e^{\wedge}(i\pi) + 1 = 0$.

Now we will see the entry of the greatest equation ever written. There are five symbols, 0, 1, e, π, i, that we frequently use in mathematics. Incredible as it might seem, they can be related by a single equation. You can practice using the **2nd** key for e, π and i by entering

$$\text{2nd\_e}^x \;\; \text{2nd\_}i \;\; \text{2nd\_}\pi \;\;) \; + \; 1$$

Getting around: navigation and editing keys

We all make mistakes; correcting them on a graphing calculator is relatively easy. The basic navigation device is the blue arrow wheel. Your right thumb does all the arrow key work.

Correction keys: ←, 2nd_INS, ♦_DEL and CLEAR

When you entered an expression and pressed **ENTER**, the input remained highlighted on the entry line. Pressing any key other than an arrow will delete a highlighted expression. But pressing the left or right arrow will unhighlight the entry line and allow you to edit the expression from the beginning or end, respectively. Insert is the default mode for corrections and the cursor shows as a blinking vertical line. This means that any key you press will insert the corresponding symbol at that point in the expression. Press the backspace key (←) to delete a character to the left of the cursor. Press **2nd_INS** and the cursor will toggle to a blinking square indicating that the correction mode is now overwrite. Press ♦_DEL to delete one character to the right of the cursor. Use **CLEAR** to delete all the characters to the right of the cursor. Pressing **CLEAR** twice will clear the whole line.

➤ *Tip: To clear the entire history area, press the blue* `F1` *key and then* `8:Clear Home`*. (Pull-down menus such as* `F1` *will be explained later.)*

Deep recall: `2nd_ENTRY`

Often we see errors to correct after entering a line or we would like to incorporate previous results into a new expression. We can access not only the previous entry line, but anything in the history area, even entries that have scrolled off the screen.

For example, the pitch of a musical note is determined by the frequency of its vibration (measured in hertz). Middle C vibrates at 263 hertz. The frequency of a note n octaves above middle C (use negative numbers for octaves below) is given by $V = 263 \cdot 2^n$. To find the frequency two octaves higher, we enter $263 \cdot 2^2$ (see Figure 1.4). To find the frequency two octaves below, we can edit the first entry and insert a negative sign before the 2.

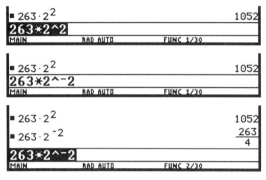

Figure 1.4 Using the recall entry and the insert mode to find two similar calculations. (Three partial screens are shown.)

In addition to having the last line available, by repeatedly pressing `2nd_ENTRY` key, all the previous entry lines in the history area are accessible. This is called deep recall. This method erases anything on the entry line and cannot be used for inserting — it is all or nothing. However, to insert a result (or a combination of results) from the history area, use the up arrow which takes you to the history area and press `ENTER` to insert the highlighted expression into the entry line. And finally (as if we need yet another way), this calculator supports the common word processing commands for cutting (◆_X), copying (◆_C) and pasting (◆_V) a highlighted expression. To highlight any set of characters, hold down the shift key (↑) and arrow over the desired characters.

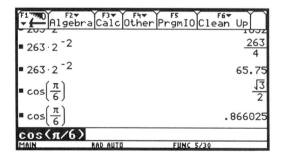

Figure 1.5 Using ◆_≈ to convert fractions to decimals and exact answers to approximations.

8 PART I / PRECALCULUS

The format of numbers

Most calculators show decimal results, but the TI-92 gives exact answers whenever possible. This can sometimes be inconvenient and you may want to change the format of the answer.

Fractions and decimals: ◆_≈

You may have been surprised by 263/4 being shown as an answer in Figure 1.4. In Figure 1.5 the history area shows how we have continued from Figure 1.4 by pressing ◆_≈. This gives us the same answer but in decimal form. Next we see that cos(π/6) returns a closed form (exact) answer, but again, pressing ◆_≈ translates the answer to a decimal approximation.

➤ *Tip: Although the green symbol ≈ only appears above one* **ENTER** *key, all three have the same effect when pressed after* ◆.

Scientific notation: Folding a paper to reach the moon

If you use really big numbers, they will be displayed in full accuracy. Let's try an unrealistic but surprising situation that uses a big number.

If you fold a piece of paper, it will double in thickness. You can measure the thickness in sheets: one fold has a thickness of 2 sheets, two folds 4 sheets, three folds 8 sheets, etc. The formula for doubling is $S = 2^n$, where n is the number of folds and S is the number of sheets. Verify the astonishing fact that it takes only 42 folds to reach the moon from earth. (We should mention, though, that it is physically impossible to fold a single sheet more than about seven or eight times.) Enter 2^42 and find the answer in number of sheets in an exact integer form. See Figure 1.6. But long strings of digits are often hard to immediately see as billions or trillions. Press ◆_≈ and the answer will be shown in scientific form. The exponent 12 gives us the magnitude. We have an answer in trillions (4.39805E12 means 4.39805×10^{12}). Notice that the exact answer appears to be rounded to six significant digits but, in fact, all the decimal accuracy is retained, as we see in the next step.

Now to verify that this stack reaches the moon, we will convert this number of sheets to miles. On the entry line of Figure 1.6, you see the expression ans(1): this was not typed, but was automatically pasted there when we pressed the divide key at the

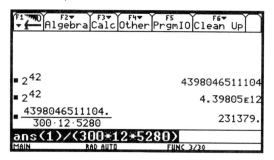

Figure 1.6 Folds to the moon. Large numbers can be forced into scientific notation.

start of the calculation. If you start an expression by pressing an operation key (+, -, *, /, ^), the calculator assumes the first number in the calculation is the previous answer. We divide our number of sheets by 300 (an approximate number of sheets per inch) and by 12 (inches per foot) and by 5280 (feet per mile). Be certain to put this product in parentheses so the calculator knows what you intend as the denominator. This conversion to miles shows that the thickness of the folded paper is more than the distance from the earth to the moon, about 230,000 miles. (Part III Demonstration B explores the unit conversions built into the TI-92 Plus.)

To insert `ans(1)` anywhere in an expression, press `2nd_ANS`. Using `ans(1)` insures that full accuracy is used, even if the previous answer was shown in an approximate format. The `(1)` identifies the first answer above the entry line in the history area. Thus, `ans(2)` is the second answer above the entry line, etc.

If you want to enter five billion without entering all those zeros, you can use the `2nd_EE` key to insert the E symbol. Press

5 2nd_EE 9 ENTER

This result is not shown here, but will be displayed as `5.E9`. Note that pressing `2nd_EE` results in a single E being shown on the screen.

➤ *Tip: In the history area of the screen, a number whose absolute value is less than 0.001 is displayed in scientific notation unless the default settings are changed.*

How to put yourself in a good MODE

The `MODE` setting choices are described in the TI-92 *Guidebook*, pages 35 and 36. We will mention other settings as we need them, but unless noted otherwise, all our examples will assume that the default settings are in effect.

Numerical format

You can control the output format of numeric calculations so that they are all shown in scientific notation. Or, if you are doing a business application, you might want money answers to come out rounded to two decimal places for the dollar and cents format. The `MODE` key allows you to check and change formats. See Figure 1.7.

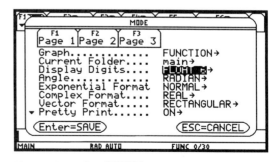

Figure 1.7 The `MODE` *screen. (The TI-92 without Plus does not have a* `Page 3`.)

To change a setting, use the down arrow to reach the desired line, then use the right arrow to display the submenu of settings. Arrow to the one you want and press ENTER to make the selection. Pressing ENTER again returns you to the HOME screen and saves your changes. Pressing ESC exits from either a submenu or the main menu without making any change.

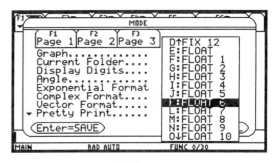

Figure 1.8 The MODE *screen and the* Display Digits *submenu.*

Status line indicators

The status line shows several MODE settings. We mention here those that we have seen so far. At the far left is the name MAIN: this is the current folder being used. The TI-92 has the capacity of a small computer and folders help organize stored files. Since we are just getting started, we will keep all our work in MAIN.

The next indicator location to the right is usually blank; it shows whether a modifier key is in effect with one of the three symbols discussed above. Then there is the indicator RAD, showing that angles are measured in radians (this is discussed more fully in the next section). The AUTO indicator means calculations are displayed in exact form where possible unless there is a decimal in the input expression. The next indicator is the graph mode; in the beginning we will graph functions so FUNC should be shown. Finally you see the history count, telling you the number of stored history pairs and the current limit on history.

You will occasionally see a busy indicator in the far right when the TI-92 is working on a long calculation.

Are your angles in radians or degrees?

Since you are in calculus, you will normally use the RADIAN setting. For situations where degrees are specified, you can change the MODE setting to DEGREE. However, it is recommended that you always leave the calculator in RADIAN mode and use 2nd_D to paste the degree symbol into the calculation where degrees are used (see Figure 1.9). You can always check the angle mode by looking for either RAD or DEG on the status line.

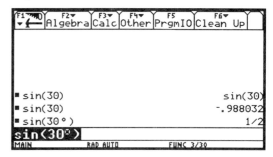

Figure 1.9 Using the degree symbol while in RADIAN *mode.*

➤ *Tip: Many handy characters, such as the degree symbol °, are unlisted on the keyboard. To see a map of the 2nd key options, press ♦_K.*

➤ *Tip: If your output values are in an unexpected or undesirable format, check the MODE settings. If you are having trouble changing MODE settings, you may have forgotten to press ENTER the second time to save new settings.*

Stored values and symbolic variables

In Figure 1.10, the store key STO→, which appears on the screen as →, is used to save a numeric value into a letter variable. Variable names can be up to eight characters in length, but a name cannot start with a number. For example, if you wanted to repeatedly use the area of a 10 inch radius pizza in calculations, you would enter

Figure 1.10 Using STO→ to store a variable

π * 1 0 ^ 2 STO→ a ENTER

The variable a can now be used in computations, as shown in Figure 1.10, where we first find the difference between the areas of 10 and 12 inch pizzas and then the cost per square inch of a 10 inch radius pizza that costs $12.99.

A symbolic processing example: factor

Although algebra is not a topic of this book, we will show a simple example of using the TI-92 for the algebraic manipulation of factoring. Many high school students suffer frustration from this topic, but the TI-92 never complains — it just does it!

Let's start with three simple quadratic expressions where we know the factoring. The first quadratic factors over the integers, the second over the real numbers and the third over the complex numbers. In Figure 1.11 two different forms of factor are applied to our three quadratics.

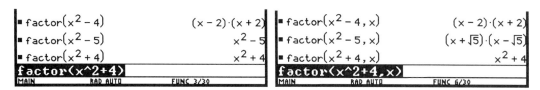

Figure 1.11 factor *(expression)* uses integers while factor *(expression, variable)* uses all real numbers.

In Figure 1.12 the complex form `cFactor` is used. When the TI-92 cannot factor the expression with the specified constraints, its "answer" is the unsimplified quadratic.

The TI-92 operates using real numbers unless you specify that you want complex answers. From this example you should be able to guess that the two pairs `solve` and `cSolve`, `zeros` and `cZeros` operate in the same manner; the leading `c` indicates complex answers are allowed. We will consider a similar problem in Chapter 6.

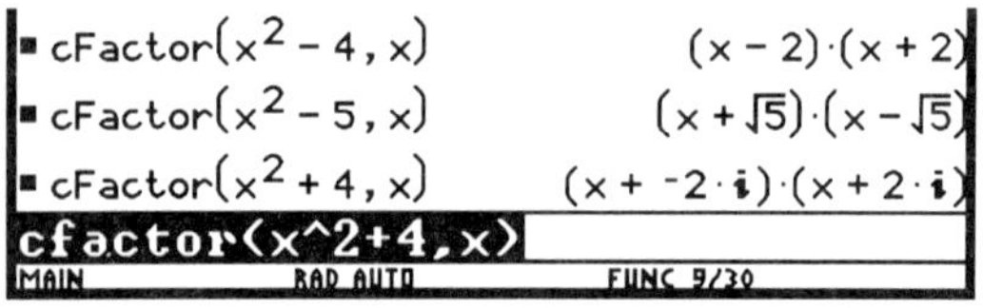

Figure 1.12 `cFactor`*(expression, variable) uses complex numbers (which include the reals).*

➤ *Tip: The TI-92 is not case sensitive, so typing* `cfactor` *instead of* `cFactor` *makes no difference. Regardless of how it is entered, the command will appear in the history area in its standard form.*

Defined and undefined variables

We end this chapter with a note about the most frequent source of error in symbolic processing. Variables are either defined or undefined. They should be undefined when used symbolically. For example, in the last example, x was an undefined variable in the expressions. If it were defined, i.e., if x were storing a number, then the expression would be evaluated with the stored number in place of x. See Figure 1.13 where x is defined to be 5 and the factor command works numerically on the expression ($5^2 - 4$) which is 21. Thus `factor(21)` gives the prime factorization 3·7.

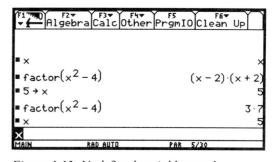

Figure 1.13 Undefined variables work symbolically, defined variables work numerically.

There is no way to know if a variable is defined or undefined by its screen representation. If you want to check a variable's status, type its name and press **ENTER**: if it shows as a number, it is defined. See the different response on the top and bottom line in the history area of Figure 1.13. An alternate but less convenient way of checking on a variable is `2nd_Var-Link`; see page 331 of the TI-92 *Guidebook*.

Cleaning up and clearing variables

The pull-down menu F6 (Clean Up) has two options for clearing variables on the TI-92 Plus. The first, 1:Clear a-z, is the only option on the TI-92 without the Plus chip. A dialog box is used to confirm this action. By using single letters for defined variables, you can quickly clear them all by using F6 before you start a new problem.

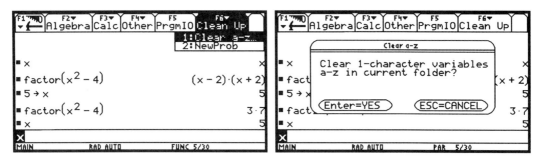

Figure 1.14 Use F6 *(Clean Up)* 1:Clear a-z, *to clear single letter variables.*

The second option, 2:NewProb, will paste NewProb on the entry line and you confirm this action when you press ENTER. So what is the difference? NewProb clears one-letter variables, but also clears the history area. It does a few other things like deselecting functions which we will find out about in the next few chapters.

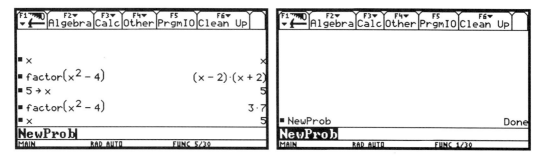

Figure 1.15 Use F6 *(Clean Up)* 2:NewProb *to clear single letter variables and the history area..*

➤ *Tip: Be sure the command* NewProb *is pasted on a clean entry line. If it is appended to any existing command on the entry line, an error message will appear.*

➤ *Tip: Longer named variables can be deleted by using the command* DelVar *or through the* 2nd_Var-Link *menu.*

2. FUNCTIONS: A FUNDAMENTAL TOOL FOR CALCULUS

The definition of functions and the use of functional notation are vital to success in calculus. In the next three chapters, we will use the TI calculator to define and evaluate functions, to make tables of values, and to graph functions. In short, we will see how to view functions analytically, numerically and graphically (the Rule of Three). In the next three chapters we will use the top row (just under the screen) of the QWERTY, to activate the graph and table features. We first focus on ♦_Y= (the green mode on the letter W) to define functions.

Formula vs. function notation: using the Y= editor

Function notation is used in calculus, whereas formulas are used in algebra. So what is the difference and how are they related? They both express a relationship between variables. Let's take the famous formula for the area of a circle, $A = \pi r^2$. In precalculus you learned to write this in functional notation $f(r) = \pi r^2$. The functional notation tells you <u>explicitly</u> which variable is the independent variable.

You can define up to ninety-nine functions in your calculator by using the function editor; the editing screen appears when you press the y(x)= key. The available functions are labeled y1, y2, ..., y99. To define a function, it is easiest to think of it in formula form. Let one of the y's be the dependent variable and make x the independent variable. For example, in our circle area formula we would define y1=πx^2, where r, the independent variable, is replaced by x.

Functions can be defined from the entry line using the keypad; see the examples in Figure 2.1. Notice that the entry line shows the keystrokes to input the function, but the definition area shows the pretty print form.

You can type a function definition directly onto the entry line, but some functions, such as the absolute value, can be pasted into a function definition from a menu. Press 2nd_MATH 1:Number 2:abs(ENTER to paste the notation in place at y4= and finish by pressing x) as shown in Figure 2.2. It should be noted that the menu system has numbered items which can be selected by pressing the appropriate numbers. Alternatively, as you become familiar with the menus, you can use the arrow keys to navigate and highlight the options, then press ENTER to select. We will learn more about the menu system as we go.

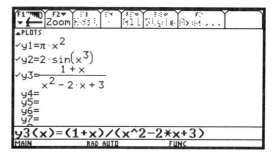

Figure 2.1 Defining functions from the keypad in the Y= editor.

2. FUNCTIONS: A FUNDAMENTAL TOOL FOR CALCULUS   15

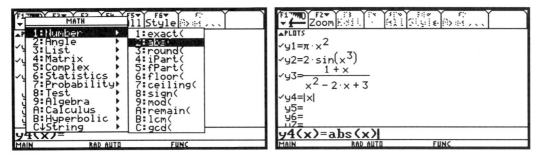

*Figure 2.2 Pasting an expression from the **2nd_MATH** menu to define **abs(x)**.*

➤ *Tip: Like some key symbols, menu item such as **abs(** will appear differently on the screen from the menu.*

Pasting from the CATALOG

Remembering the spelling and location of special symbols and functions within various menus can be tedious. This is why there is a built-in alphabetic listing of all functions and settings on the calculator. If you don't know the menus well, then the most convenient way to paste an expression into a function definition is to use the **2nd_CATALOG** key (the yellow manifestation of **2**). In the catalog screen you can move quickly to the function you want by pressing the letter key that starts its name. For example, if you want to use the hyperbolic tangent (tanh), then press **2nd_CATALOG** and you will see the list beginning with your last entry you viewed in the catalog. Now press **T** (unless you are already there). Finally arrow down to highlight the entry **tanh(** and press **ENTER**.

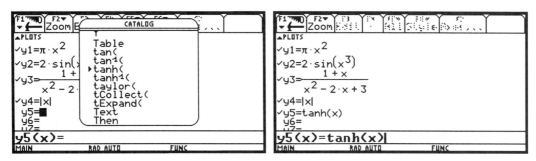

*Figure 2.3 Using **2nd_CATALOG** to define a function from the catalog. Press **T** and arrow down to the desired item. Pressing **ENTER** pastes the expression to the cursor position on the entry line.*

➤ *Tip: When in the catalog (although not shown in Figure 2.3), the status line will show the syntax needed to complete the command. The command syntax for this example is **tanh(EXPR)** where EXPR means an expression.*

Cleaning up and getting out: ESC or 2nd_QUIT

On the function editor screen, just as for the HOME screen, use the arrow keys to navigate. Use DEL and INS to edit. Pressing CLEAR will delete any definition. Use ESC to back out of menus or use ♦_HOME (or 2nd_QUIT) to directly return to the HOME screen.

➤ *Tip: Be careful about using the CLEAR key. It instantly deletes the entire entry and there is no recovery other than re-entering the formula.*

A summary of menu use

There are several types of menus. The first is a pull-down menu from the tool bar at top of the screen. In the first screen of Figure 2.4 you see the F3 (Calc) menu which gives quick access to useful commands in calculus. The F1 (Tools) menu is also shown and has some special visual cues that enhance the menu system. Notice that the first entry is dotted which means that it is not available for use in this context. The down arrow has been used to highlight an item. If we press ENTER we can expect a dialog box to be presented because of the ellipsis (...) symbol in its name. Another feature is that ♦_F is shown as a key equivalent command; this means that ♦_F can access this dialog box directly from the HOME screen to avoid using the pull-down menu. You will find that, once you have become familiar with the menus, pressing the number (or letter) of the item will be preferable.

In a dialog box, you will either have a box to fill from the keyboard or a list to choose from (indicated by →). To keep the settings you have made in a dialog box you must press ENTER to save them. If you press ESC, they will revert to the previous settings.

➤ *Tip: A common mistake is to change settings and go directly to another menu (like pressing ♦_HOME) which will not save the changes to the settings. It is a good habit to always exit all dialog boxes by using either ENTER or ESC.*

In Figure 2.1 we encountered the massive MATH menu. All of its items are submenus and are shown with a pointer arrow (▸) to alert you that this is a submenu. The right arrow (or ENTER) will open a submenu. You can exit layers of a menu with the ESC key.

In summary, look for the special symbols that will help you use menus efficiently: submenu ▸, dialog box ... and option list →.

2. FUNCTIONS: A FUNDAMENTAL TOOL FOR CALCULUS   17

▶ *Tip: A quick way to know if you have a TI-92 Plus is to see if* `F6` *is labeled* `Clean Up`. *On the TI-92 without the chip, the label is* `Clear a-z`.

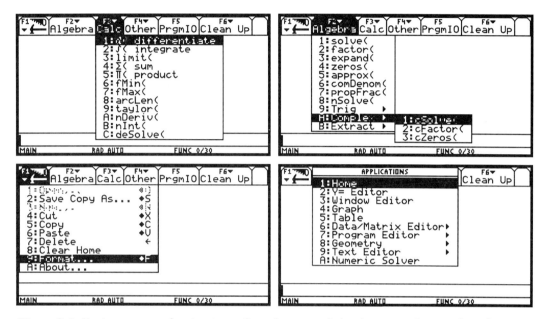

Figure 2.4 *Various menus showing icons for submenu* ▸, *dialog box* ... *and option lists* →.

You can guide your work by using the large blue `APPS` (applications) key. However, the first five menu items are accessible by using the ♦_QWERY keys. Until we have use for the other applications, we will not reference this menu.

In summary, we have seen that menu choices will either paste an item to the cursor location, bring up a new menu line, or bring up a new screen.

Evaluating a function at a point

A benefit of functional notation $f(x)$ is that $f(3)$ is conveniently understood to be the output value of the function when 3 is input. We have entered a function `y1` using the `Y=` editor, but from the `HOME` screen we can define functions with other names. For example, we use the `Define` command to create an area function, `f(r)=πr^2`. (Rather than type `Define`, paste from `F4` (`Other`) `1:Define`.) To find the area of a circle with radius 10 cm, we simply enter `f(10)`.

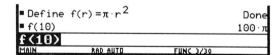

Figure 2.5 *Evaluating* `f(10)` *on the home screen. (Partial screen shown.)*

New functions from old

In the next two examples, we create new functions from previously defined ones.

Composite functions: f(g(x))

Suppose an oil spill expands in a perfect circle and that the radius increases as a linear function of time. We can create a new composite function that expresses the area in terms of time. Let $f(r) = \pi r^2$ and $g(t) = 1+t$, where t is in hours. Define a new composite function named c as $c(t) = f(g(t)) = \pi(1+t)^2$. Notice in Figure 2.6 that c(t) gives the composite definition in terms of t. To find the area of an oil spill after 2 hours enter c(2); this gives the same answer as f(g(2)) (using ans(1) = g(2)).

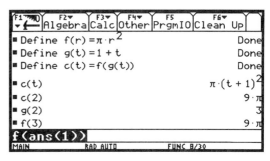

Figure 2.6 *Creating a composite function for area in terms of time.*

A Malthusian example

In 1798, Thomas Malthus proposed that population growth was exponential and that food supply would grow at a linear rate. We return to the Y= editor where we model a food supply (per million persons) as y1 = 5 + .2x; this means that there is food for five million people in the base year and that each year afterwards the supply increases to provide for an additional 200,000 people. (As you enter these functions, the previous ones will be erased.) For the population (in millions), set y2 = 2(1.03)^x; this corresponds to an initial population of two million that increases annually by three percent. Let y3 = y1(x) − y2(x) and y4 = y1(x) / y2(x). These two new functions are a measure of excess food and a measure of food per capita, respectively. In Figure 2.7 these two measures are evaluated at 50 and 100 years from the base year. There is a shortage in the hundredth year (i.e., y3 is negative and y4 is less than 1).

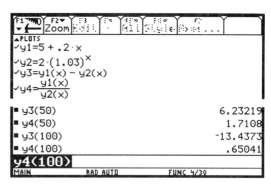

Figure 2.7 *Food excess and food per capita in 50 and 100 years. (Partial screens:* Y= *on top,* HOME *on bottom.)*

Defining families of functions by using lists: y1={...}x

Let's see how to define a family of linear functions in just one single function definition. Suppose a city has three taxi companies. Red charges $1.00 to get in one of its taxis and $0.40 for each eighth of a mile traveled. Green charges $2.00

2. FUNCTIONS: A FUNDAMENTAL TOOL FOR CALCULUS

to get in and $0.30 for each eighth of a mile traveled. Blue charges $3.00 to get in and $0.20 for each eighth of a mile traveled. The cost in terms of miles traveled is linear; for example, Red has the cost function $C = 1 + 3.2x$ (where x is in miles, so the coefficient is $8 \cdot 0.4 = 3.2$). We can model this linear situation, $C = b + ax$, with just one function definition whose parameters, a and b, are lists.

We use the STO→ key to enter the function definition from the HOME screen in Figure 2.8. (Along with Define and the Y= editor, this makes three different ways to define a function.) We have redisplayed the definition on the entry line to point out how to enter lists. Remember to use braces ('curly brackets'), the 2nd version of parentheses. And be certain to separate list entries with commas — you see that list entries are separated by spaces in the pretty print of the history area, but entering values that way will give the product.

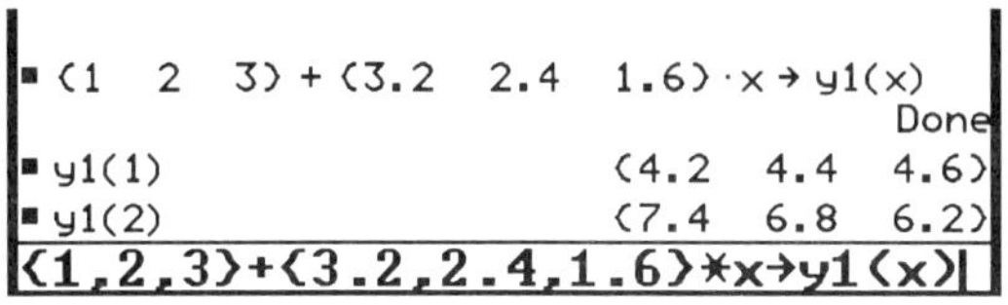

Figure 2.8 A family of functions created using a list and evaluated simultaneously.

The function evaluations shown give a list of the fares for one and two mile trips. We see that Red is cheapest for a one mile trip but Blue is cheapest for a two mile trip. It should be clear from the rate structure that Blue will be cheapest for longer trips.

A note about defining functions on a TI-92 without Plus

When a function is defined on the TI-92 without the Plus chip, the variable name itself can not be used later in an input expression for the function. (Unless the input is just that variable, in which case the definition appears.) Consider the simple example x^2→f(x). If you enter f(a+1) the result will be (a+1)², but if you enter f(x+1) a Circular Definition error message is displayed. To avoid this problem, most users define functions with a repeated letter variable name, such as xx^2→f(xx). In this book, the functions have been defined on a TI-92 Plus where this special restriction is unnecessary. Thus, if you are using a TI-92 without the Plus chip and you encounter a Circular definition error, just redefine the function using repeated letters for variable names.

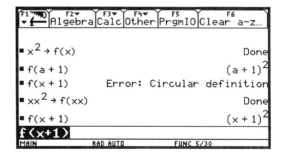

Figure 2.9 The Circular definition error on a TI-92 without the Plus chip and a way of avoiding it.

3. TABLES OF FUNCTION VALUES

This chapter will focus on making lists and tables of function values. These values often reveal the nature of the functional relationship: are the numbers increasing? decreasing? periodic?

Lists of function values: f({...})

To see a set of values for a function, you can evaluate a function with a list as the input variable and output a corresponding list of function values. For example, suppose you want to find the area of circles with radii 10 cm, 50 cm, 100 cm, 200 cm and 500 cm. Define the area function and evaluate the function with a list as shown in Figure 3.1. Use the left/right arrow keys to scroll horizontally when data is too long to fit on the screen. Look for ▸ at the right or ◂ at the left of the screen to indicate that there are more values off the screen in that direction.

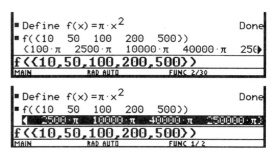

Figure 3.1 The values of the area function found as a set {10, 50, 100, 200 and 500}. Use arrows to scroll the answer set.

A table of values for a function: TABLE

There is a much more convenient way to see a list of value for a function, but we must use one of the reserved Y= functions. In the Y= editor, press F1 8:Clear Functions and then ENTER, before we start. We simply enter the function, y1=πx^2, set the beginning and increment for a table, and display it. To set up a table, press ♦_TblSet and complete the dialog box. Let's redo the above problem by using a table beginning at zero (tblStart=0) and having the *x*-values go up by ten (△tbl=10; remember to press ENTER twice). Now press ♦_TABLE to display the table; see Figure 3.2. The values beyond 70 are not shown on the

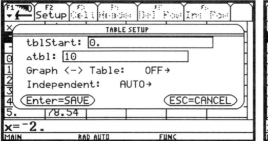

Figure 3.2 Making a table for the area function y1=πx².

first screen, but we can use the down arrow to see them. You can also use F2 (**Setup**) to return to the table setup dialog box and adjust the values.

Selected values for a table: Independent: Auto Ask

Suppose we just want the function evaluated at the specific list of values from above. In table setup, we can arrow down to the option Independent: Auto and use the right arrow to change from 1:Auto to 2:Ask. Now press ENTER to save the change and press ♦_TABLE. The previous table values will appear, but you can write over previous *x*-values and customize the list. See Figure 3.3. Some of the values are listed in scientific notation, but by highlighting the cell you will see the full stored value.

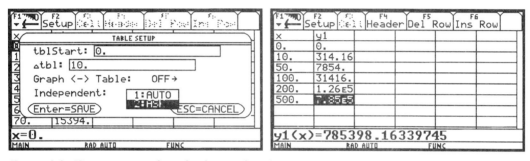

Figure 3.3 How to enter selected values rather than an incremented list.

➤ *Tip: You can use ♦_F (same as F1 9:FORMATS) to adjust cell width.*

➤ *Tip: If some of the rows already show values you want to include, you can use F5 (Del Row) to delete the ones you do not want and then add other values.*

A table for selected functions

Recall the food supply and population functions from the Malthus model of the previous chapter. If we want to enter these again but also want to keep our area function as y1, then we can use y2 and y3 in the Y= editor; see Figure 3.4. If you

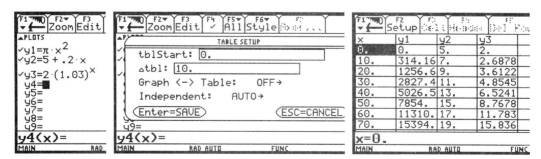

Figure 3.4 Using a table for selected functions. (Partial screens.)

Selecting and deselecting a function

When you want to see values for only certain functions, you can deselect the ones you do not want and select the ones you do want. In the Y= editor, move the cursor to function definition line and press F4 (✓). This is a toggle: if it was checked, it will turn off; if it wasn't checked, it will turn on. When you make a function definition, it is automatically turned on. To avoid having y1 displayed, deselect y1, as shown in Figure 3.5, so that only y2 and y3 show in the table display. Recall that F2 (Setup) is a convenient access to reset the table settings.

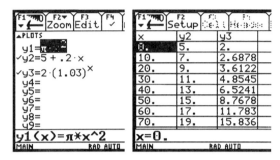

Figure 3.5 With y1 deselected it will not appear in the table.

Find the zero of a function from a table

A natural question arises from the Malthus model: When will the food supply no longer be sufficient for the population? In the previous chapter, we had a function that measured the excess food supply, so we want to find the years for which the excess food function gives negative values. However, it is easier to start by looking for when the excess food function is zero. Enter the relevant functions again and turn off any functions that we don't want to see. Since we know from before that the excess is positive at year 50 and is negative by year 100, we set tblStart=50 and ∆tbl=10 (see Figure 3.6). We see that the zero is between year 70 and 80. Now use ♦_TblSet and set tblStart=75 and ∆tbl=1.

We could continue searching for a more precise value by setting tblStart=79 and ∆tbl=0.1.

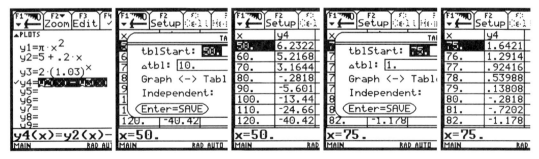

Figure 3.6 Searching for a zero of the food excess function. (Partial screens.)

> *Tip: Deselected functions are still active for calculations when used in other function definitions.*

Editing a function formula from inside a table

You can redefine a function from the table itself. Suppose you want to look at the food ratio function in the Malthus model. See Figure 3.6. Arrow over to the y4 column and press **F4 (Header)** to put its equation on the entry line of the screen. Press any arrow to un-highlight the equation, replace − with ÷, then press **ENTER** to finish editing.

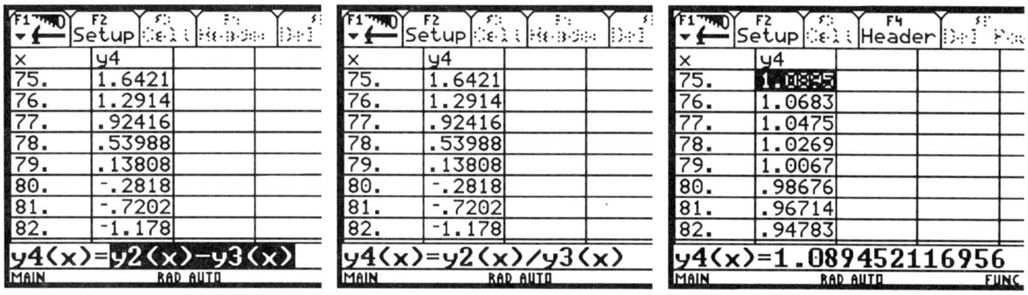

Figure 3.6 Changing a function definition from inside a table by using **F4 (Header)**.

4. GRAPHING A FUNCTION

This chapter will continue our investigation of the graph/table features accessed from the QWERTY keys and show how to graph the functions that we have defined.

Basic graphing: ♦_Y=, ♦_WINDOW and ♦_GRAPH

Graphing is like the 1-2-3 of taking a picture with a camera.

- *Select your subject(s).* To select a function, recall that you use F4 (✓) from the Y= editor. (Deselect or clear functions that you do not want to graph.)
- *Frame them properly.* Press ♦_WINDOW and set the *x*- and *y*-window boundaries.
- *Click to take the picture.* Press ♦_GRAPH.

The hard part of photography is getting the subject both in the picture and looking good. On the calculator we control the picture by using the WINDOW menu. In Figure 4.1 the first row shows the settings and graph for a picture of the function y1=πx². The function's graph uses too little of the screen. The second row in Figure 4.1 shows an improvement made by changing two settings in the WINDOW screen.

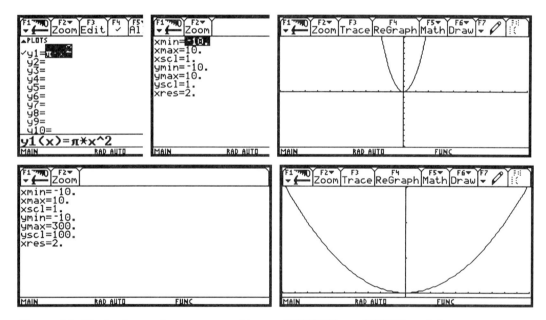

Figure 4.1 The basic graph sequence: ♦_Y=, ♦_WINDOW, ♦_GRAPH *and then an improvement by changing the* WINDOW *settings.*

➤ *Tip: Pressing* ♦_HOME *will take you back to the* HOME *screen.*

4. GRAPHING A FUNCTION 25

Window settings: xmin, xmax, xscl, ymin, ymax, yscl, xres

The window setting variables are as follows:
xmin sets the left edge of the window as measured on the horizontal axis,
xmax sets the right edge of the window as measured on the horizontal axis,
xscl (*x*-scale) sets the width between tick marks on the horizontal axis,
ymin sets the bottom edge of the window as measured on the vertical axis,
ymax sets the top edge of the window as measured on the vertical axis,
yscl sets the width between tick marks on the vertical axis, and
xres (*x*-resolution) sets the selection density of values to plot (1 is the highest setting and 2 is the default which should be used unless the graphing is very slow).

Finding a good window: Zoom

Like photography, the setting of the window is an art. It is rare for us to know the ideal window before we graph; trial-and-error experimentation is usually required. To help with this process, we can use the F2 (Zoom) menu to get started.

With thirteen choices (notice the submenu), it can be hard to remember them all. Let's begin with three items that address a common need — a quick window setting.

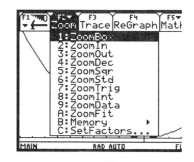

Figure 4.2 The Zoom *menu.*

Zoom special settings: ZoomStd, ZoomDec, ZoomTrig

There are three Zoom menu options that automatically set the window to special settings and graph the selected functions, all in one keystroke. These special settings are shown in Figure 4.3. They are especially

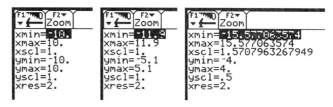

Figure 4.3 ZoomStd, ZoomDec, *and* ZoomTrig *window settings.*

helpful when you are graphing common functions (such as polynomial, exponential, and trigonometric) with graphs close to the origin. The 6:ZoomStd (Zoom Standard) setting often works well as a good first view. The 4:ZoomDec (Zoom Decimal) option gives what is called a *nice* window because the *x*-values used to graph progress from -11.9 to 11.9 by tenths. The nicety of this will be explained in the tracing section just ahead. For trigonometric functions, the obvious first choice for graphing is 7:ZoomTrig (Zoom Trigonometry); it uses *x*-values from -(119/24)π to (119/24)π (shown in decimal form; think of it as just shy of -5π to 5π). Using ZoomStd will reset xres to 2, but the other two will not change the xres setting.

Window adjustment for Malthus: ZoomFit

When graphing functions that model a situation, you will almost always know the domain of the function, but probably not the range. If you have entered the domain, you can use A:ZoomFit, another Zoom option, to help look for the best *y*-values. Let's try this out on the Malthus model of the previous chapters.

Malthus never published a graph; he used only numerical and analytical expositions. Some of his readers didn't see his concern. A graph appeals to a broader audience. We will now find a good window for the Malthus graphs. Enter or turn on the two Malthus equations. Because we want to look from the base year to a century in the future, we will set xmin=0 and xmax=100. Now the range of the functions is difficult to guess, so we leave the setting in ZoomStd (i.e., ymin=-10 and ymax=10). See Figure 4.4.

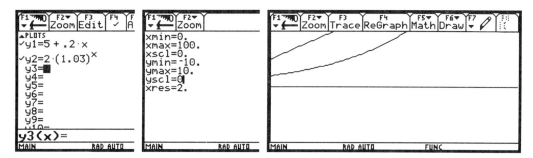

Figure 4.4 Using Y=, WINDOW, *and* GRAPH *to get a first graph with a good domain.*

➤ *Tip: Set* xscl *and* yscl *to zero as you adjust a window, then choose helpful values for them when the final window size is found.*

The window does not show the population growth for all the domain, so in Figure 4.5 we use F2 (Zoom) A:ZoomFit to reset the *y*-axis window settings so that all the *y*-values of the function will be shown. (The *x*-axis window settings are unchanged by ZoomFit.) To see the values that

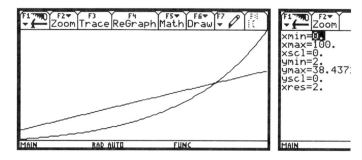

Figure 4.5 Using ZoomFit *to find the range.*

have been set, press ♦_WINDOW. From these values you make a final decision to use xscl=10 and yscl=10 that will help us read approximate values from the graph.

Since there are no numeric labels on the graph, it is often necessary to alternately use the ♦_GRAPH and ♦_WINDOW keys to check on the window dimensions. This problem of knowing where you are on an unmarked graph is the next topic.

Identifying points on the screen

By pressing any arrow key with a graph screen showing, a cross-hair cursor as shown in Figure 4.6 will appear in the center of the screen and the *x*- and *y*-values of the cursor point are displayed. This cursor is called the free-moving cursor. You can use the arrow keys to navigate this cursor to any point on the screen. By placing the free-moving cursor on the graph of a function, you can display

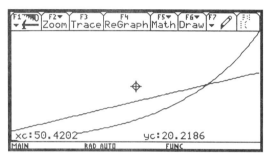

Figure 4.6 The free-moving cursor, activated by an arrow key and deactivated by the ENTER or CLEAR key.

the approximate coordinate values of the point $(x, f(x))$. The *x*- and *y*-values are labeled xc and yc. If you press ENTER or CLEAR, the free-moving cursor mode is canceled.

Reading function values from the graph: Trace

You will be more interested in the values of the function than general points on the screen. We improve on the free-moving cursor approach by using the Trace mode. Pressing F3 (Trace) evokes the trace cursor. The trace mode displays the *x*- and *y*-values of the current trace cursor location. You see in Figure 4.7 that

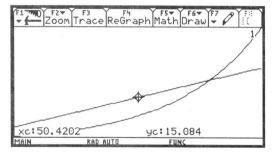

Figure 4.7 The Trace cursor on a function.

the number of the graph being traced is displayed in the upper right corner.

In Figure 4.8 we first use the right arrow to move to the right on the graphs shown in Figure 4.7. You can also enter a specific *x*-value: from the TRACE mode, just type the desired number and press ENTER. (Pressing F3 again will move to the nearest point normally displayed on the graph.) In the second panel of Figure 4.8, the trace cursor is switched to the other graph: the up and down arrows select the function to be traced. In this case, we started on y2 and went to y3 with the same *x*-value. Pressing CLEAR cancels the Trace mode.

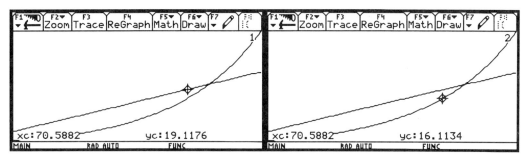

Figure 4.8 Right/left arrows move along the graph. Up/down arrows move between selected graphs.

Making better trace values: ZoomInt

As you arrow right (or left) using Trace, you see that the *x*-values have long decimal expansions. There is little need to know the food supply after exactly 70.5882 years. If we just wanted a rough idea about values, we could ignore the extra decimal digits.

In Figure 4.9 we use 8:ZoomInt to reset the window so that the trace values will be integers. You are first asked for the desired center of the graph and we move the cursor to xc=50.4202 and yc=20.2186 before pressing ENTER. Now use Trace to see that the *x*-values are all integers. Use ♦_WINDOW to see the adjusted settings.

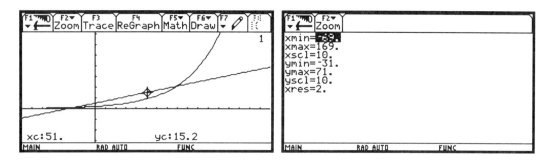

Figure 4.9 ZoomInt sets the x-values to integer values for nice tracing.

How to make a nice window (an optional adventure)

The screen is 239 pixels wide and this is why the `ZoomDec` setting is so nice: with `xmin=-11.9` and `xmax=11.9`, the resulting *x*-values are 119 negative tenths, zero, and 119 positive tenths, a total of 239 pixels. With `ZoomInt` in the last example, the *x*-values started at `xmin=-69` and ended at `xmax=169`, 239 values altogether (counting zero). The same holds for `ZoomTrig` which spans 239 intervals of length π/24. For the Malthus graph, if we wanted a nice window that started at `xmin=0`, then we could set `xmax=238`. In general, for a graph starting at `xmin=0`, set `xmax=23.8*n`, where *n* is large enough to let 23.8·*n* span the *x*-values you want to include. This produces a nice window!

➤ *Tip: Don't confuse* `8:ZoomInt` *with* `2:ZoomIn`.

Panning a window

There is an old story about blindfolded people describing an elephant from different perspectives; their guesses included wall, tree, and snake. Sometimes functions are like elephants: you may need to take the blindfold off to see the whole picture. In case the subject is too big to fit in one window, we can move the window frame to see what is to the left or the right or above or below the current view. This is called panning. Let's take an example of a logistic equation and start as if we knew nothing about it.

$$f(x) = \frac{1000}{1 + 9e^{-0.05x}}.$$

Enter the function as `y1=1000/(1+9e^(-.05x))` and follow the steps below to practice panning a window.

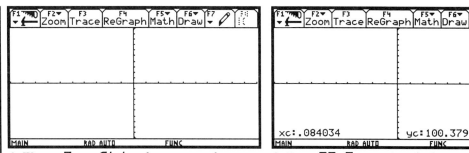

1. We use `ZoomStd` *and see no graph. Recall that the range is* $-10 \leq y \leq 10$.

2. Press `F3` *(*`Trace`*) to find* <u>some</u> *point and see that* `ymax=10` *is too small.*

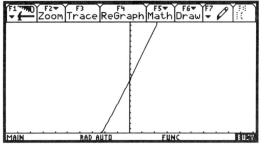

*3. Press **ENTER** to 'quick zoom.' This pans the window and places the trace cursor at the center. In this case, the view is too small (roughly $-10 \leq x \leq 10$ and $90 \leq y \leq 110$).*

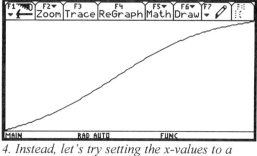

*4. Instead, let's try setting the x-values to a broader interval. Set $0 \leq x \leq 100$ and use **ZoomFit**.*

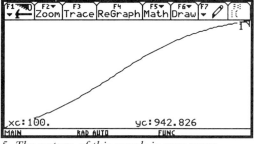

*5. The nature of this graph is now more evident. Use **Trace 100** give the function value at the right edge of the window.*

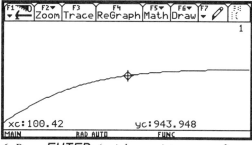

*6. Press **ENTER** (quick zoom) to pan and move the point $x = 100$ to the center of the graph.*

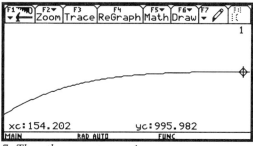

*7. The other way to pan is to attempt to trace beyond the right side of the screen; the screen will pan to the right. Use **2nd_**arrow keys to move trace faster.*

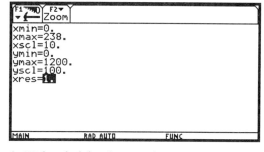

*8. With a feel for this graph now, we use nice integer boundaries so that when we trace we will see integer values and set **xres=1**.*

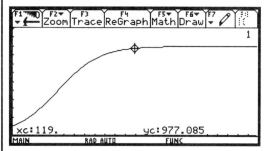

*9. Press **F3** (**Trace**) to see the final view.*

Figure 4.10 Panning a window

So far, we have used two techniques to move a window without giving specific numbers. First is the quick zoom command, Trace ENTER, which pans and centers the window on the trace cursor. Second is moving the trace cursor to the left of the xmin value or to the right of the xmax value which horizontally pans the window.

➤ *Tip: Pressing 2nd_arrow keys will put the trace cursor in <u>turbo</u> mode — it moves faster.*

➤ *Tip: Unlike the trace cursor, the free-moving cursor will stop at the edge and does not pan the window.*

What is a good window?

You have seen the trial-and-error approach to finding a good window, but this begs the question: what is a good window? For a function serving as a model of a physical situation, a good window will show you the function's graph for the relevant domain. For example, there would be no interest in negative values of the area function $A = f(x) = \pi x^2$. But considering the same function as a purely algebraic quadratic function, we would like to choose a window that includes negative x-values so we can see all of its important behavior. Whenever possible, we want to show asymptotic behavior. For example, we needed to see past $x = 100$ in the graph of the logistic function because that function is approaching the line $y = 1000$. We call this the end behavior. However, if we concentrate solely on the end behavior, we might blur some local behavior. In the logistic example, an important local behavior is the point of inflection where the graph changes from concave up to concave down. (Concavity is discussed in Chapter 10.)

Can you always find a good window?

No. There are pathological functions that we cannot graph and others that require than one view to show all their important behavior. This case occurs when the end behavior view makes it impossible to see the local behavior and vice versa.

Asymptotic dangers: Beware of graphs with vertical lines

Sometimes the graphing calculator will lead you astray. The most common case is rational functions. Let's take the blind graph approach to

$$p(x) = \frac{x^2 + 2x + 30}{x - 4}$$

First, we enter y1=(X²+2X+30)/(X-4) (don't forget the parentheses) and use `ZoomStd` to get some idea about the function. We need to see a bigger picture. We now use `3:ZoomOut`. This will display the small cross-hair cursor to designate the center of the expanded window to come. The current center (the origin) is OK, so press `ENTER`. The result is a window with the domain and range both four times as big.

In Figure 4.11 we are left with an unexplained vertical line to the right of the origin. Could this be part of the graph? A careful look at the function will tell you that the denominator is undefined at $x = 4$. Recall from precalculus that this line is called an asymptote. But the calculator did not draw it as an asymptote. The calculator draws graphs by connecting special x-values that are found by starting at `xmin` and adding increments of (`xmax` - `xmin`) / 239. In this case, to the left of $x = 4$, $f(3.69748) = -168.803$, and to the right of $x = 4$, $f(4.36975) = 156.415$. (You can verify these values using `Trace`.) Connecting these values gave us the vertical line.

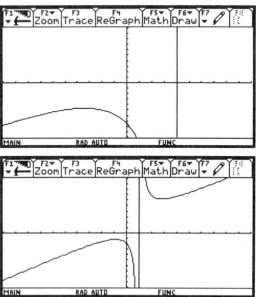

Figure 4.11 Graphing a rational function.

One way to prevent connection across an undefined point is to create window settings with the undefined value of x exactly in the middle of `xmin` and `xmax`. See Figure 4.12. The trace cursor 'shows' that the value is undefined at $x = 4$ by not showing a y-value there.

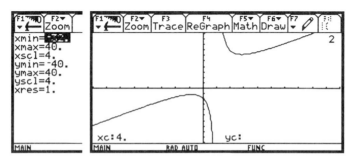

Figure 4.12 Window setting so that the trace cursor shows no y-value at the undefined point $x = 4$.

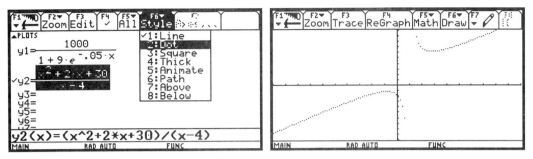

Figure 4.13 Changing the graph format to dot display.

Changing plot style

An easier way to remedy the connecting problem is to change the style mode in the Y= editor from 1:Line to 2:Dot in the F6 (Style) menu as shown in Figure 4.13. The other styles are:

- 3: square dots
- 4: bold connected line
- 5: round cursor without path
- 6: round cursor with path
- 7: shading above
- 8: shading below

An inaccurate graph

A rational function graphed on a calculator may connect dots when it should not. A similar distortion occurs when the resolution is insufficient to display a function. Consider

$$f(x) = \sin\left(\frac{1}{x}\right)$$

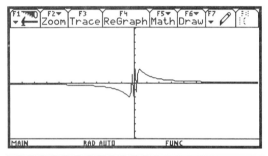

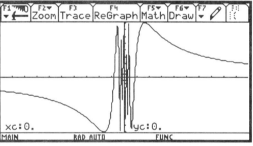

There is no way this function can be accurately graphed if the origin is shown. Several attempts are shown in Figure 4.14. The xc:0. and yc:0. display in the second screen is misleading because the function is undefined at 0; it is a residue from the ZoomIn prompt for a center. The third screen correctly indicates that y has no values when $x = 0$.

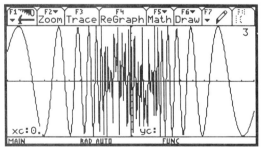

Figure 4.14 Three attempts to graph y1 = sin(1/x) accurately using the window settings ZoomTrig, ZoomIn, and $-.1 \leq x \leq .1$.

➤ *Tip: In writing mathematics, it is good style to write decimal values less than one with a leading zero. Therefore, in the caption of Figure 4.14, we should have written $-0.1 \leq x \leq 0.1$. However, when `0.1` is entered in the calculator, it is converted and shown as `.1`. For this reason, we will often break with style and not use a leading zero.*

Other Zoom options

For the sake of completeness, we will mention the other choices on the `Zoom` menu. The `1:ZoomBox` option is similar to `ZoomIn` and `ZoomOut` in that it displays the small cross hair cursor on the graph. Move this cursor with the arrow keys to the screen location where you want a new window to have one corner, press **ENTER**, then move the cursor to the diagonal corner desired (you will see the rectangle on the screen as you move the cursor), and press **ENTER** again. The new window will be the rectangle you defined.

The `B:Memory` submenu includes the `1:ZoomPrev` option which returns you to the previous settings. This is handy if you have changed the window settings and the graph is worse. You can also keep one setting in memory: store a window setting with `2:ZoomSto` and recall it with `3:ZoomRcl`.

The `5:ZoomSqr` (zoom square) selection is helpful for graphing circles. It is similar to `ZoomFit` in that it is a one-keystroke grapher that fixes one axis. Here the x- and y-values are adjusted so that the axes have the same physical scale on the screen (i.e., so that circles look like circles, not ellipses).

The `9:ZoomData` entry graphs a good window for statistical data. We will use this only once in a later chapter.

Finally, `C:SetFactors...` (set zoom factors) leads to a dialog box that allows you to change the setting for multiplier/dividers `xFact`, `yFact` and `zFact` used in `ZoomIn` and `ZoomOut` (`zFact` is used in three-dimensional graphing.) The default setting of 4 is good, but sometimes 2 is convenient.

5. EXTENDED GRAPHING FEATURES

We have seen how to set up and graph a function. In additional, we have used F2 (Zoom) and F3 (Trace) from the graph screen. In this chapter we will see how to use some features of the other pull-down menus from the graph screen.

The most important menu is F5 (Math). In later chapters it will be used frequently to access the calculus items that are on this menu. In this chapter we will restrict our attention to the Math menu items that are used to identify some special points on a graph. The chapter continues with instructions on split screen graphing, drawing inverses, and placing text on graphs.

➤ *Tip: This chapter's references to* Math *are all to the graph screen's menu item F5 (*Math*). This should not be confused with* 2nd_MATH *above the 5 key.*

Finding roots: Zero

We first use the Math tool 2:Zero to find a zero of a function. We return to the Malthus model (y1 and y2 as shown below) and consider the food excess function y3=y1(x)-y2(x). The zero of this function has an important meaning: it is when we start having a food shortage.

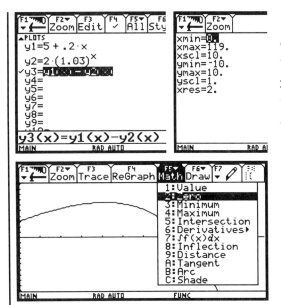

Although not essential, we first turn off y1 and y2. We use the ZoomStd window but make xmin=0 and xmax=119 to better show the excess function graph. Now press ♦_GRAPH to see the graph.

From the GRAPH screen press F5 (Math) and select 2:Zero. This pull-down menu contains the most commonly used items in precalculus and calculus.

36 PART I / PRECALCULUS

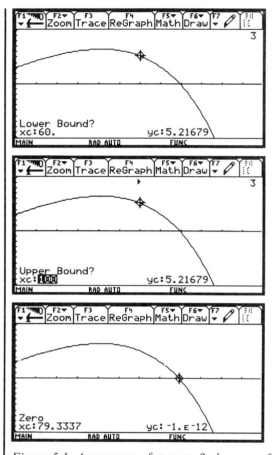

We are prompted for a lower bound. Since we are to the left of our zero, we just press **ENTER**. Optionally, you can enter an *x*-value.

After pressing **ENTER**, a right facing arrow on the screen will mark the lower bound and you will be prompted to enter an upper bound. Enter an *x*-value large enough, or arrow to the right far enough, so that the function values are negative there. (We enter 100.)

Press **ENTER** and the root is shown graphically and given numerically at the bottom of the screen. So why isn't yc:0? The exponent (E-12) makes the *y*-value close enough to zero for the calculator.

Figure 5.1 A sequence of steps to find a zero of a function.

➤ *Tip: The special points of the* Math *menu can only be found between the current* xmin *and* xmax *settings.*

➤ *Tip: The closer the bounds, the faster the TI finds the zero.*

Finding extrema: Minimum and Maximum

The sequence of steps is the same whether you are finding a zero, a maximum, or a minimum. On the F2 (Math) menu, both 3:Minimum and 4:Maximum use the same request for a lower bound and an upper bound.

For the excess food function, we might ask in what year the excess is a maximum. By using Trace on the graph in Figure 5.2, we can see that the graph is displayed as a horizontal line segment between years 36 and 47. This is misleading because the value of the function is not a constant on this interval. Rather, the resolution of the calculator screen is limited. We use 4:Maximum to

5. EXTENDED GRAPHING FEATURES   37

find the highest value. Again we enter a lower bound and an upper bound as numeric estimates or by using the arrows

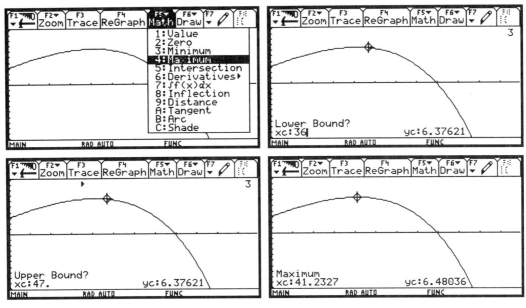

Figure 5.2 Finding the maximum of a function within an interval.

Finding an intersection of two graphs

A typical task is to find where two functions are equal. Analytically, this means finding the zero of the difference function as we have done. Graphically, this means finding where the graphs of the two functions intersect.

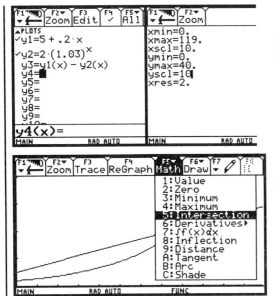

We reset the window for $0 \leq y \leq 40$ and select y1 and y2, the food and population functions.

From the F5 (Math) menu select 5:Intersection.

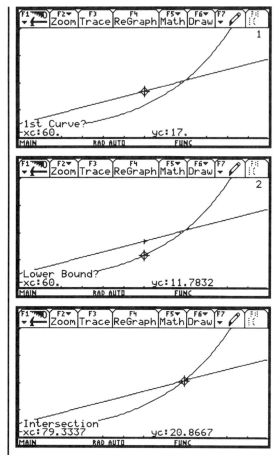

You are asked to identify the first curve; press **ENTER** to accept the default choice (identified by the flashing trace cursor). Select the second curve by pressing **ENTER**. If more than two curves are graphed, choose by using the up or down arrows.

Now the calculator prompts you for a lower bound and an upper bound; these must enclose the desired intersection point. (This is necessary since some graphs have multiple points of intersection.)

We see that the intersection point is at 79.3337 years. We got the same answer as we did in Figure 5.1 using **Zero** on the difference function — we must have done it right!

Figure 5.3 Finding an intersection requires choosing two curves and a lower and an upper bound.

Split screen graphing (optional)

The screen of the TI-92 is quite wide in comparison to its height. There are occasions when it can be helpful to see two screens, say a graph and table, at the same time. As another example, showing the **WINDOW** screen beside the graph can help interpret graph points. The split screen option is on page 2 of the **MODE** menu. There are three different settings for **Split Screen**:
```
1:FULL
2:TOP-BOTTOM
3:LEFT-RIGHT
```
We will only use 1 and 3. Move the cursor to 3 and press **ENTER** twice.

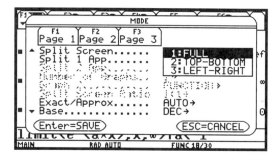

*Figure 5.4 Split Screen options in **MODE** (**Page 2**). (The TI-92 has no **Page 3**.)*

To move from one half-screen to the other, press `2nd_APPS` (corresponding to the yellow icon of a split screen with arrows).

The easiest way to exit the split-screen mode is to press `2nd_QUIT` from the entry line. This means that if you are not on the entry line, you will need to use `2nd_QUIT` twice in succession. You could also `MODE`.

➤ *Tip: Notice that you see the same graph using a split screen, but with a different ratio in the x-values. Therefore some* `Zoom` *options will set a different window when there is a split screen.*

Graphing inverse functions

We now use a split screen to graph inverse functions. Recall that for any function f, the graph of f^{-1} is the reflection of the graph of f about the line $y = x$. This can be seen in Figure 5.5 for the function $f(x) = x^3$. The window has been set by using `ZoomDec`. To distinguish the different graphs, we use Style options. From the `Y=` screen, highlight the desired function, press `F6` (`Style`), and select the style. In Figure 5.5, we used `4:Thick` for $y2=x^{1/3}$ and `2:Dot` for $y3=x$ (the default is `1:Line`, the style for $y1=x^3$).

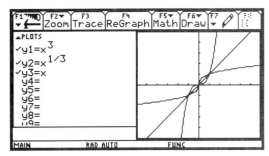 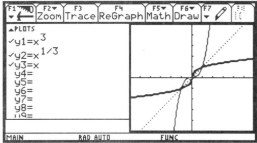

Figure 5.5 *A function and its inverse in a split screen with the* `ZoomDec` *setting, then a change of styles from the* `Y=` *editor and a better graph.*

We now graph $f(x) = \sin(x)$ and its inverse, $\sin^{-1}(x)$, in a `ZoomDec` window. We see in Figure 5.6 that some of the graph's reflection across the $y = x$ line is missing. The difference from the previous example with the cubic function is that the reflection of sine is not a function. The domain of $\sin^{-1}(x)$ has to be restricted to make it a function.

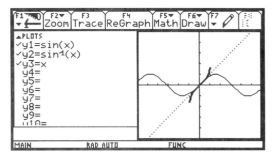

Figure 5.6 *A function and its inverse where the inverse is not a complete reflection across the line* $y = x$.

To graph a full reflection of the function, we use the **F6** (**Draw**) pull-down menu from the **GRAPH** window and select **3:DrawInv**. This pastes **DrawInv** on the entry line of the home screen (with its history showing) and we add the function so that the full command is

   DrInv sin(x)

Pressing **ENTER** will replace the home screen with the graph screen as shown in the last frame of Figure 5.7. You can see why restrictions are necessary on the domain of $\sin^{-1}(x)$.

Putting text on graphs

It is possible to put text on a graph screen using **F7** (✎) and **7:Text**. Move the cross hair cursor to the location where you want the text and type. You can enter characters from the keyboard or entries from menus, but you cannot paste items from the **CATALOG** or use **2nd_RCL** to paste equations. See Figure 5.8.

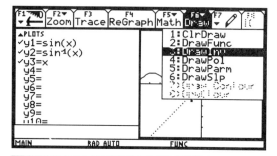

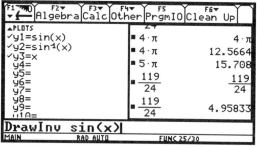

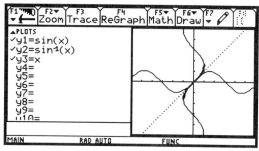

Figure 5.7 The **3:DrawInv** command will draw a complete reflection even though the graph is not a function.

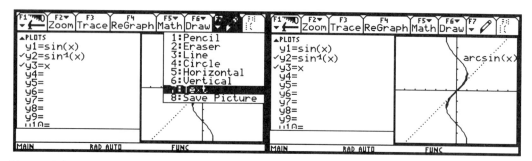

Figure 5.8 Putting text on a graph.

6. SOLVING EQUATIONS

Mathematicians don't use many verbs. They mostly say "this equals that" and shout a few commands like find, evaluate, simplify, and solve. What does it mean to *solve* an equation? For the equation $3x + 2 = 0$, the single solution is $x = -2/3$. The quadratic equation $x^2 - 2x - 3 = 0$ has two solutions, $x = 3$ and $x = -1$, as can be seen by factoring $x^2 - 2x - 3$ to $(x - 3)(x + 1)$. In these examples, there is just one variable and we want to know all values of the variable that make the equation true. If we switch to function notation and let $p(x) = x^2 - 2x - 3$, then solving the quadratic equation is the same as finding the zeros of the function $p(x)$. In the last chapters we showed how to graph functions and, among other things, find their zeros. A more direct approach without graphics is available using, in the F2 (Algebra) menu, either 1:solve or 4:zeros.

This will be our first thorough investigation into the computer algebra system of the TI-92. The calculator is capable of working symbolically with variables, giving function answers in addition to numeric data. This power brings with it a requirement for careful syntax. Learn the commands carefully; it's worth the effort.

Also, the first notable differences between the TI-92 without and with the Plus arise in this chapter. Their output screens will differ in some cases; remember that the screens in this book come from the TI-92 Plus. The TI-92 Plus has the added capability of the Numeric Solver that we will use in the last section.

Solving single variable equations

We start be looking at a typical situation such as solving the quadratic equation mentioned above.

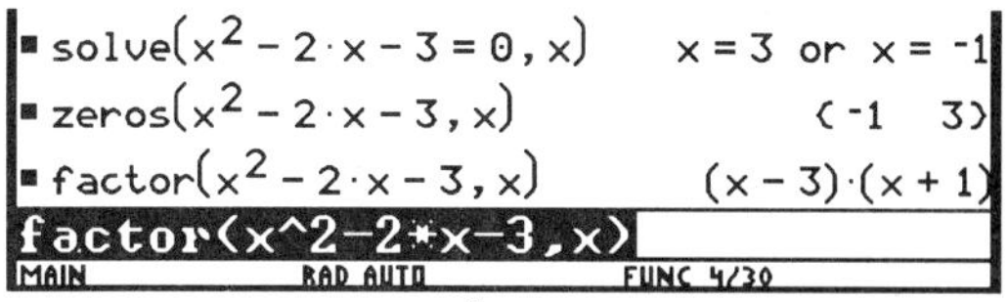

Figure 6.1 Using solve, zeros *and* factor *with a particular quadratic expression.*

Solving a quadratic equation

You must give solve two things, the equation you want to solve and the name of the independent variable. The command zeros is a slightly faster way to solve equations with zero on one side; the calculator actually invokes the solve command to execute zeros. The syntax for zeros is the input expression whose zeros you want (not an equation) and the name of the independent variable. For many purposes, the output of zeros is in a more convenient form. For example, the set of zeros can be the input for a function and evaluated. In Figure 6.1 we see a comparison of these two commands, followed by the factor command which is related to finding the zeros of an expression.

The calculator can perform algebra with symbolic parameters. So, for instance, it should be able to give us the famous quadratic formula. Our first effort toward this end, shown in Figure 6.2, just spits back our input equation (the TI-92 without Plus gives the even more frustrating response `false`). The calculator always reports back with whatever simplifications and solutions it can do — in this case, it wasn't what we hoped for. The problem is with how we entered the equation. Look carefully: there is no multiplication dot in the history area between `a` and `x`². We entered `ax`² and the calculator understood that to be the two-letter variable `ax` squared. Despite our intentions, we entered an equation in three variables, `ax`, `bx` and `c`, and asked it to solve in terms of another variable, `x`, not present in the equation. No wonder the TI couldn't do anything! The same entry with `a*x`² and `b*x` gives much better results. (Rather than re-typing, edit the previous entry by inserting two `*` symbols.) The answer is not in the traditional form that you may have memorized, but it is clearly equivalent. Chapter 19, Application 2 (page 344) of the TI-92 *Guidebook* goes through the "completing the square" derivation of the quadratic formula in a display of the calculator's symbolic algebra capabilities using `expand` and `factor`.

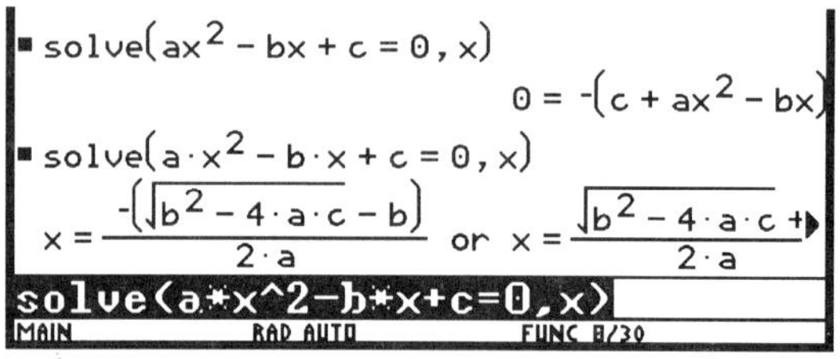

Figure 6.2 Trying to find the quadratic formula (the answer is too long to fit on the screen).

A common problem with symbolic computations is that the TI may have a longer memory than you do. If you did not get the same responses shown in Figure 6.2, then perhaps you gave one of the letters we used a value. Even if it was days ago that you typed `c=5` or `a→3`, the TI remembers and dutifully evaluates each instance of the variable with the value you assigned to it. Use `F6` from the `HOME` screen to clear all one-letter variables and avoid such problems. Some longer variable names are reserved words and must be avoided (see page 491 of the *Guidebook*).

➤ *Tip: Any time you enter an expression or equation in a command, you must identify the independent variable. The* `Too few arguments` *error message will remind you to do this should you forget.*

➤ *Tip: The first entry for* `solve` *is an equation while the first entry for* `zeros` *or* `factor` *is an expression (no =).*

Solving a quadratic equation with complex solutions: cSolve

When we try to solve the quadratic equation $x^2 + 5 = 0$ in Figure 6.3, the result is a terse `false`. This means that the equation has no real solutions. A response of `true` is given if the equation is an identity, i.e., it is true for all reals. (The TI 92-Plus is better at recognizing symbolic identities than the TI-92 without the chip.) Trying to factor $x^2 + 5$ leads nowhere, but we know that this equation has complex number solutions. As we saw in Chapter 1, the F2 (Algebra) menu has a submenu A:Complex with complex number commands. Using cSolve gives the complex number solutions of the quadratic equation and cFactor works over the complex numbers, not just the reals, giving the factorization into linear terms guaranteed by the Fundamental Theorem of Algebra. The appendix of this book gives more information about using complex numbers on the calculator.

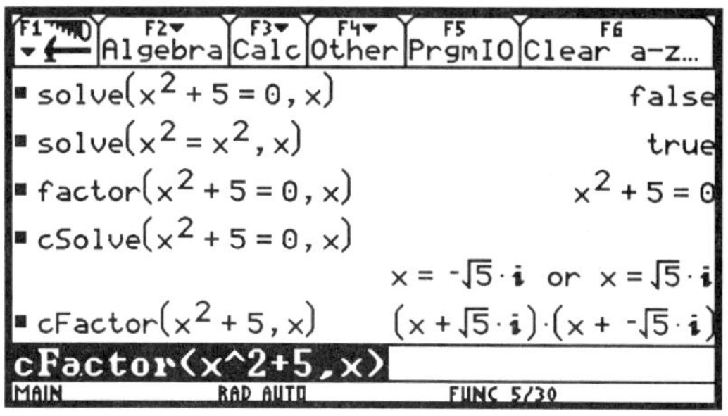

Figure 6.3 Complex analogs of solve and factor from the F2 submenu A:Complex.

▶ *Tip: If you find* solve *or* factor *is inadequate and you want to use* cSolve *or* cFactor, *then just add* c *in front of the command. It is not necessary to capitalize* S *and* F.

Solving non-polynomial equations

In the preceding examples, we could use algebraic techniques to reach a solution. High school algebra is sometime taught as though it can be used to solve any kind of mathematical problem. Yet there are many equations, even some polynomials (of fifth and higher degree), that have no algebraic means of solution. But they do have numerical solutions. For example, consider four equations in Figure 6.4, each sets the basic exponential function equal to a simple linear expression.

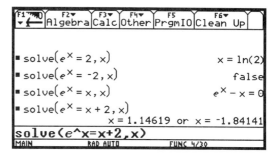

Figure 6.4 Several types of output from solve.

The symbolic solution to $e^x = 2$ is derived by taking the inverse of both sides: it is an exact answer. (Be sure when you enter the exponential function that you use **2nd_e**^{x} above the **LN** key, not the letter **e** that would be read as a variable.)

44 PART I / PRECALCULUS

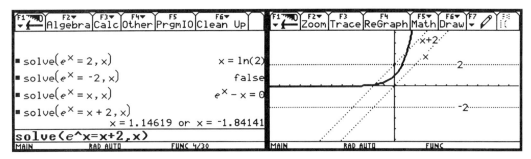

Figure 6.5 Several types of output from `solve` *and the graphic representation of looking for intersections. The* `Thick` *style function is* e^x *and the equations' right-hand sides are drawn with the* `Dot` *style and labeled. (The left screen repeats Figure 6.4.)*

Next we look for the point where the exponential function equals -2, but of course this never occurs and the response `false` means that there is no real solution. This agrees with the graph; the exponential curve never intersects the lower horizontal line. But even though it does not intersect $y = x$ either, the calculator's response to the third `solve` command is a difference equation. This is another instance of the TI doing what it can and reporting back something other than what we wanted. (The TI-92 without the Plus responds with `false`, which is preferable here.) The last case gives two numeric answers. Realize that these are approximations; there is no way to use an inverse and get an exact solution as we did in the first case.

The graph in Figure 6.5 shows $y = e^x$ as a thick line and the four lines in the dotted style (remember `F6` (`Style`) in the `Y=` screen). We have also added text labels to each line.

▶ *Tip: The entry* `cSolve(e^(x)=-2,x)` *(not shown) doesn't give* `false`*, rather an answer involving the square root of -1 and the symbol* `@`*, explained below. You would need to take a class in complex analysis to fully understand this, but it is noteworthy that the TI gives the correct answer for complex-valued logarithms.*

Equations with an infinite number of solutions

How does `solve` handle periodic functions? If an equation has infinitely many answers, then `solve` will respond with all of them. We consider a simple case, sin(x)=1/2. In the first entry of Figure 6.6, we use .5 for 1/2 which results in a decimal answer. (When the `Exact/Approx` mode is set to `AUTO`, a decimal in the input

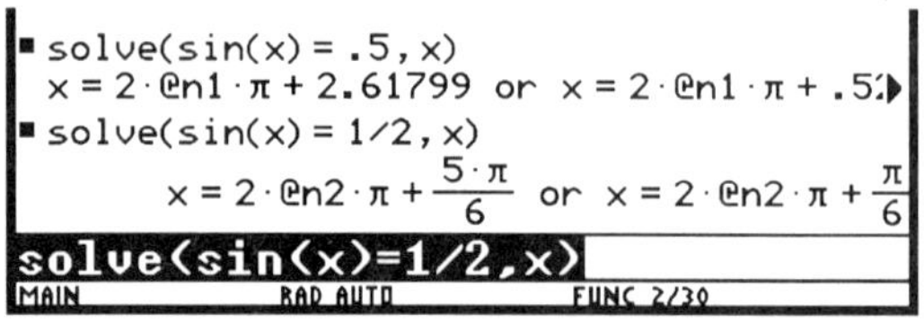

Figure 6.6 `Solve` *answers when there are infinitely many solutions.*

will cause numeric answers to be decimal approximations.) Repeating the command with the fraction 1/2 gives the answer in exact form. The symbols @n1 and @n2 signify any integer. The second answer is often written as $5\pi/6 + 2n\pi$. Until we clear variables, successive answers of this sort will have the symbols @n3, @n4, etc.

➤ *Tip: The infinite answer variables are most easily reset on the TI-92 Plus by using the F6 (Clean Up) option 2:NewProb.*

Conditional solutions

A powerful feature of the TI-92 is the ability to impose conditions on solutions. It is rather simple. For example, suppose we want only positive answers. In Figure 6.7, at the end of the first `solve`, command we add the vertical bar symbol "|"

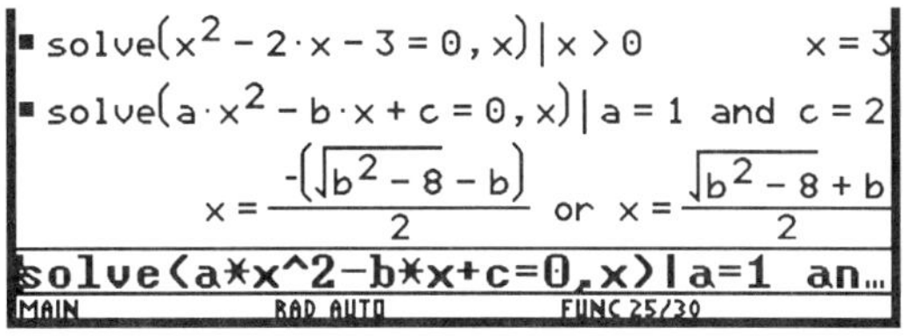

Figure 6.7 Using `solve` *with conditions.*

(located above the K key and read "such that" or "with") and the condition x>0. Another use of this technique is that you can enter an equation with letter parameters and define any known parameters as a condition. These definitions are not permanent, but only used for this entry. More than one conditions require the conjunction `and`.

Simultaneous solutions to systems of equations

With a TI-92 Plus, the `solve` command works with simultaneous solutions to systems of equations. Use `and` to list the equations and make the variables to solve for a set. In Figure 6.8 we solve a simple system of linear equations and then find the intersection points of the unit circle with a parabola. Notice the use of `and` and `or` in the solution.

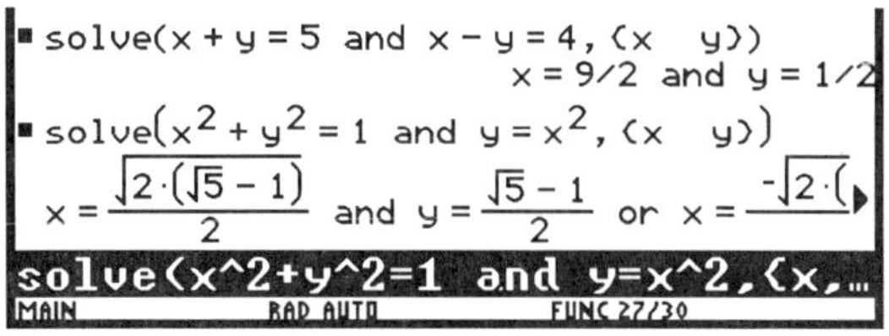

Figure 6.8 Using `solve` *with simultaneous systems of equations.*

What if? analysis of an investment using Numerical Solver

Equations commonly have several variables and we have seen that `solve` can handle these because we specify the variable to solve for. The TI-92 Plus has the option A:Numeric Solver in the APPS menu which is handier than `solve`; this is the analog of Solver from earlier TI calculators. (The APPS key is just below and left of the arrow key.) For example, the formula for calculating the growth on a continuously compounded investment (a type of exponential growth) is given by

$$P = P_0 e^{kt},$$

where P is the future worth, P_0 is the present value, k is the rate of return, and t is the time (in years) of the investment.

Typically in an investment opportunity, you ask any one of four questions:
1. What will my investment be worth at some future date? (Knowing the other variables, find P.)
2. What will I need to invest now in order to get a desired amount in the future? (Knowing the other variables, find P_0.)
3. What investment rate do I need in order to have a desired amount in the future? (Knowing the other variables, find k.)
4. How long will it take my investment to reach a desired amount? (Knowing the other variables, find t.)

Variable names can be from one to eight letters long, so we choose descriptive names. Let `present` stand for the initial (or present value) amount P_0, `future` for our future value, `rate` for our investment rate, and `time` for time in years. It would be nice to set the Display Digits in MODE to have money amounts in Fix 2 format, like 134.20. Unfortunately, the MODE settings do not affect the display in Numeric Solver.

Press APPS A:Numeric Solver and enter the new equation as shown below. We will then play "what if" by setting any three of the equation variables and using F2 (SOLVE) to calculate the value of the remaining variable.

► *Tip: Before you press F2 (Solve), check to see that the cursor is on the line of the variable for which you want a solution.*

Press APPS A:Numeric Solver to begin. (Again, this requires a TI-92 Plus.)

6. SOLVING EQUATIONS

When you enter the `Numeric Solver`, any previously used equation will still be there. You may need to use `CLEAR` to start a fresh equation. `F5` (`Eqns`) gives a menu of known equations.

After entering your equation, press `ENTER`. A list of your variables and their current values appears — if they are not blank, variables may have been previously defined.

Let's answer a type 1 question: What is the future value of $1000 at 6% in 10 years? To find this, enter the values we know for `present`, `rate`, and `time`. Arrow up to `future`.

Press `F2` (`Solve`). (Answer: $1822.12) If you forget to arrow up to the unknown variable, an `ERROR` message will appear. If this happens, press `ESC` and start again.

A type 2 question, has a goal of $1500 in 10 years with a 6% rate. Enter 1500 for `future`, arrow down to `present`, and press `F2` (`Solve`). You do not need to clear the value of 1000 that exist in `present`; it is treated as a guess. (Answer: $823.22)

For a type 3 question, suppose we are promised $1500 in 10 years on an initial investment of $500. What is our rate of return? (Answer: ≈11%)

Time is the unknown in a type 4 question. This specific answer tells us how long it takes $500 to triple at 6%. (Answer: ≈18.3 years.) This is a general answer in that any amount at this rate will take this long to triple.

Figure 6.9 A sequence of screens showing the use of `Numeric Solver`.

A. THE SINE ANIMATION

This is a precalulus demonstration designed to produce a simple animation to show the sine function as the height of a point moving around the unit circle. We will simultaneously graph three parametric functions. The first will be a graph of a unit circle. This is simple with the parametric equations $x = \cos(t)$ and $y = \sin(t)$. A second graph is used to show vertical displacement and a third graph shows the vertical displacement over time.

Follow the sequence of steps outlined below.

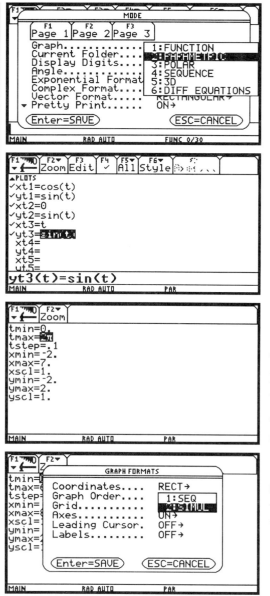

In the MODE screen, change the Graph selection to PARAMETRIC.

Use ♦_Y= and enter the three parametric function definitions shown. Notice that they are entered in pairs and t must be the variable. They are automatically selected as they are defined.

Set the window with ♦_WINDOW. The t-values go from 0 to 2π, with tstep=.1. (When you press ENTER, 2π will changes to a decimal.) The window was chosen so that the unit circle will appear circular since the axes' scales are nearly the same.

Press ♦_F and a GRAPH FORMAT screen will appear. Set Graph Order to 2:SIMUL. Now all three graphs will be drawn simultaneously.

A. THE SINE ANIMATION 49

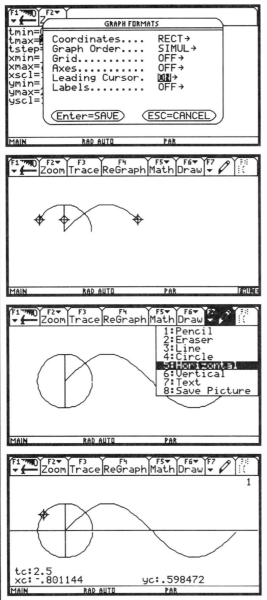

Make two more GRAPH FORMAT changes. Turn the Axes OFF; this will allow the vertical displacement graph to be seen. Also with Leading Cursor ON, all three graphs will be drawn as if the graph style were Path. (Otherwise they would have to be set individually.)

Press ♦_GRAPH and the graph will animate. Notice the motion of each of the three graphs and the fact that the three leading cursors are always at the same height. Use F4 (ReGraph) to repeat the animation.

You may also want to add the horizontal axis. This is easily done by using F7 (✐). The Horizontal line prompt will show a line at yc:0, which is where we want it so press ENTER. Press ESC to leave the line draw mode.

Use F3 (Trace) to see the coordinate values. You can read the values on another function by pressing the up arrow. On the unit circle, the value of *t* is length of the arc from the starting point to the cursor's current position.

Figure A Sequence of steps to produce the sine curve animation.

B. THE TEXT EDITOR

The `Text Editor` application is available in the menu activated by the blue `APPS` key located near the arrow wheel. The demonstration here is just an introduction of this feature. A thorough explanation and more examples can be found in Chapter 16 of the TI-92 *Guidebook*.

Setting up the Malthus model

Every TI-92 has a port with which you can connect the calculator to a view screen and project images with an overhead projector. Suppose it is your assignment to give a short talk on the Malthus model. Here is a sequence of steps that you might follow to get ready. First enter your equations and find a good window to show the graphs.

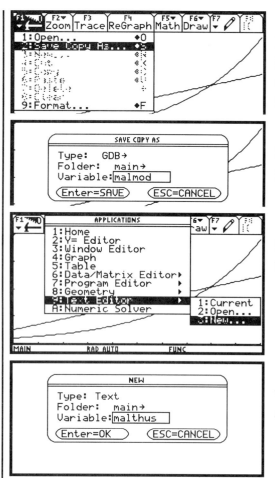

Once you have a graph that you like, use `F1` (`Tools`) and `2:Save Copy As`. This records the graph mode, the `Y=` equations, and the window for instant recall.

We name the `GDB` (Graphical Data Base) `malmod`. Press `ENTER` twice and you will be back to the graph screen.

Now we are ready to build a text file. Use the blue `APPS` key and press `9:Text Editor` then `3:New`.

We name the file `malthus`. You will then be on a blank text screen with a cursor to the right of a colon (`:`).

B. THE TEXT EDITOR   51

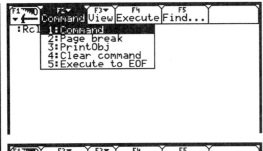

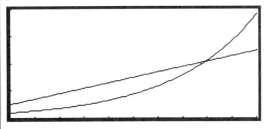

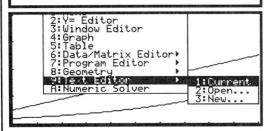

The first thing we type (or paste from the catalog) is the `RclGDB` command. Then use F2 (`Command`) 1:`Command` to place a C to the left of the colon. This means that we can later use F4 (`Execute`) on this command to recall the saved graph settings.

Now we type in the text we want to display. There is no word wrap: you would have to use spaces to push broken words to the same line. We included two other command lines that will allow us to make changes to the model functions and see a new graph.

From the `Text Editor`, place the cursor on the first line and press F4 (`Execute`). This will set up for the graph you want to show. Now press ♦_GRAPH and the graph will be shown. At this point you can TRACE or use F5 (`Math`) commands as usual.

To return to the text editor, just press APPS 9 1 and the cursor will be where it was before graphing. We can go to the population growth rate and reset it to 2%.

Use F4 (`Execute`) and ♦_GRAPH. You will see the two graphs showing no intersection. (So with 2% there is no shortage of food in the time span shown.) To return to the test editor screen, again use APPS 9 1.

The cursor returns to your place in the text file. This cursor memory is important when you have several pages of text and don't want the cursor to always return to the top of the file.

Figure B Using the `Text Editor` *for a presentation with graphs and text.*

C. COBB-DOUGLAS 3D GRAPHING

We graph a function of one variable by thinking of the input values on the *x*-axis and the output values on the *y*-axis, i.e., setting $y = f(x)$. Functions can depend on more than one variable; is there a way to visualize these? If a function has two input values, then we can think of the inputs as *x* and *y* and set a third variable $z = f(x, y)$. The resulting points $(x, y, f(x, y))$ form a surface in three-dimensional space.

The TI-92 allows us to view surfaces with the **3D** graph mode. The **Y=** editor entries become **zi=** and there are many more settings in the **WINDOW** screen. This demonstration is only an introduction to **3D** graphing.

We will consider a function from economics called the Cobb-Douglas production function. It gives a measure of productivity based on using *x* units of labor and *y* units of capital. The general function is $f(x, y) = k \cdot x^a \cdot y^b$, where *a*, *b*, *k* are positive constants with $a + b = 1$.

These **3D** graphing capabilities and the related calculus commands will be useful if you take multivariable calculus.

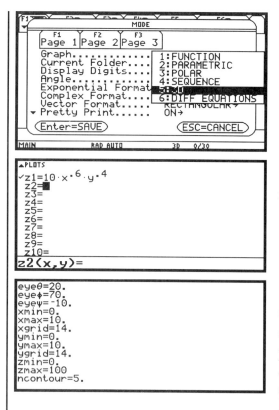

Press **MODE** and change the **Graph** setting to **5:3D**. (The TI-92 without Plus will not have a **Page 3** or the sixth **Graph** option.) Notice **3D** replaces **FUNC** on the status line.

Pressing ♦**_Y=** brings up a screen of **z** functions. Enter the Cobb-Douglas function **z1=10x^.6*y^.4**.

There are many additional **WINDOW** settings to contend with. The **eye** angles have to do with the viewing perspective of the surface. We change only the **x**, **y**, and **z** settings. (The TI-92 without Plus has only two **eye** settings and no **ncontour**.)

C. COBB-DOUGLAS 3D GRAPHING   53

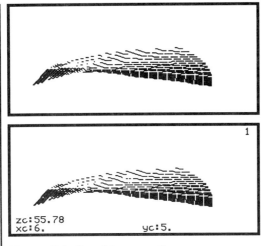

A percentage displayed in the upper left corner shows the calculator's progress computing the surface (not shown). It then takes a moment to draw the picture. You can see productivity increases as *x* and *y* increase.

You can TRACE on the surface. It is still the case that left and right arrows change the *x*-values, up and down the *y*-values, but what you see on the surface is a little different. You can also enter numeric values.

Figure C.1 Graphing a surface.

When a function has multiple inputs, you often want to compare the effects of changing one input. In this context, that question means: does increasing labor or capital have a greater impact on production? The last screen of Figure C.1 shows the point (6, 5, 55.78). Enter *x* = 5 and *y* = 6 to find *z* = 53.7827. This means that shifting one labor unit over to capital makes the productivity decline.

Another question is what inputs other than 5 and 5 give a production level of 50? A contour level is a two dimensional curve showing the points (*x*, *y*) that give the same *z*-value on the surface. It is like a line on a topographical map connecting all the points of the same elevation. The TI-92 Plus can draw contour levels.

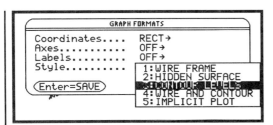

Press ♦_F to bring up the GRAPH FORMATS screen. Set STYLE to 3:CONTOUR LEVELS and then press ENTER twice. The WINDOW setting ncontour determines how many will be drawn (we use the default of 5).

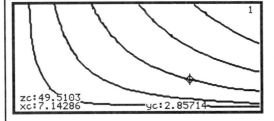

After calculation, you will see this picture. On this screen, TRACE acts like the free-moving cursor – it does not follow contour lines. We find that *x* = 7.14 and *y* = 2.86 also give a production level near 50.

Figure C.2 Showing contour levels (not available on the TI-92 without Plus).

PART II
DIFFERENTIAL CALCULUS

7. THE LIMIT CONCEPT

8. FINDING THE DERIVATIVE AT A POINT

9. THE DERIVATIVE AS A FUNCTION

10. THE SECOND DERIVATIVE: THE DERIVATIVE OF THE DERIVATIVE

11. THE RULES OF DIFFERENTIATION

12. OPTIMIZATION

DEMONSTRATIONS

   A. OPTIMIZATION WITH CABRI GEOMETRY

   B. IMPLICIT DIFFERENTIATION

7. THE LIMIT CONCEPT

A fundamental difference between precalculus and calculus is the application of the limit. In precalculus, we can define average velocity over a time period of some positive length. For example, if you drive 200 miles in four hours, then you averaged 200/4 = 50 miles per hour. But looking at the car's speedometer, you see the speed at a given moment, the instantaneous velocity.

The following table shows heights of a grapefruit thrown in the air. Find the average velocity over periods of one second. (This is quite simple, but it allows us to cover several general aspects of lists on the calculator.)

time (seconds)	0	1	2	3	4	5	6
height (feet)	6	90	142	162	150	106	30

Figure 7.1 Grapefruit height per second.

Creating data lists and finding average velocity

First we put the data in lists using the set bracket notation. This is shown in Figure 7.2. We store the two lists into variables named **time** and **height**. (Remember to use commas to separate list entries; see the end of Chapter 2 for notes on lists.) By hand, we find the average velocity for the first second:

(90 – 6) / (1 – 0) = 84

```
■ {0  1  2  3  4  5  6}→time
                {0  1  2  3  4  5  6}
■ {6  90  142  162  150  106  30}→hei▶
        {6  90  142  162  150  106  30}
  90 - 6
■ ------                             84
  1 - 0
■ Define delta(x)=seq(x[i+1]-x[i],i,1▶
                                   Done
  delta(height)
■ -------------
  delta(time)
              {84  52  20  -12  -44  -76}
```

Figure 7.2 Entering the data into two lists by hand and finding average velocity from the data. The complete **Define** *command is given below.*

The TI-92 will do this repetitive calculation and display the whole list at once. We need to define a new function **delta** that will find the difference between successive items in a list. The definition uses a very handy TI-92 function called **seq**. This is like a command from a programming language; the syntax is

seq(*expression, index, start, stop, increment* **)**

It makes a list of values using the expression which usually depends on the index variable. Index values run from the start to stop values changing by the increment value each time. (The increment can be negative; you should play with this command if you want to know more about it.) The whole definition for **delta** does not fit on a single line, but it can be read from the history area and the entry line. It is easier to press **F4 (Other)** and paste **1:Define** than to type the command. Brackets are the **2nd** version of comma and divide. The function definition is

`Define delta(x)=seq(x[i+1]-x[i],i,1,dim(x)-1,1)`

There are two more things to explain in this definition. The bracketed values refer to the entries of a list, so `x[i]` is the *i*-th entry in the list *x*. The dimension command has slightly different meanings in various contexts; here `dim(x)` gives the number of elements in the list *x* — using this in the stop value guarantees that we do not attempt to access list entries that do not exist (which would evoke an error message). We use `Delta` to find the quotient of successive differences of the two lists; division of lists is done term by term.

► *Tip: We used full screens in Part I, but from now on we will omit the pull-down menus and status line if they are not informative.*

What does the limit mean graphically and numerically?

The limit is at the foundation of calculus. For example, the key to calculating an instantaneous velocity is to let the time period become closer and closer to zero. Using data lists, this becomes virtually impossible. If we know an expression for a function, say displacement $s(t)$, then the average velocity from time a to time $a + h$ can be written

$$\frac{s(a+h) - s(a)}{h}$$

Notice that the time period h is in the denominator of the average velocity, so letting it reach zero would mean dividing by zero — certainly an error. But limits avoid that problem: the idea is to see if the function approaches some value L as the h-values get closer to zero (without ever reaching it).

The graphical approach to limits

Let's see how the limit works with two particular functions. The first function f has no denominator and is defined for all x, even at $x = 0$. It is included to illuminate the second function g, which is not defined at $x = 0$. This second function g is defined as an average velocity function with $s(t) = t^3$, $a = 1$ and $h = x$.

$$f(x) = x^2 + 1 \quad \text{and} \quad g(x) = \frac{(1+x)^3 - (1)^3}{x}$$

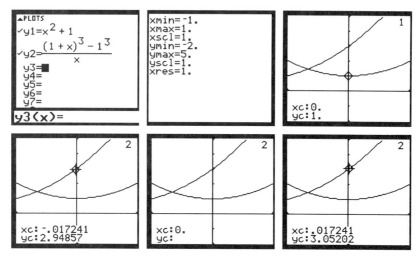

Figure 7.3 When the trace cursor is shown at x = 0 (top row), the first function is 1. The trace cursor for the second function (second row middle), however, is undefined at x = 0: the y-value is blank. Notice that y-values on either side of the y-axis are very close to 3. This is the graphical meaning of the limit. (Split-screens and cropping are used in this display.)

Now look at the *y*-values of *g*(*x*) for *x*-values on each side of zero. As shown in Figure 7.53, one is 2.9857 and the other is 3.05202; it doesn't take a rocket scientist to figure out that the values are getting close to 3.

For these two cases we write $\lim_{x \to 0} g(x) = 3$ and $\lim_{x \to 0} f(x) = 1$.

The numerical approach to limits

Now let's use the table approach in Figure 7.4 to see the same thing. Press TABLE and modify settings with TblSet (or F2 (Setup) from the TABLE

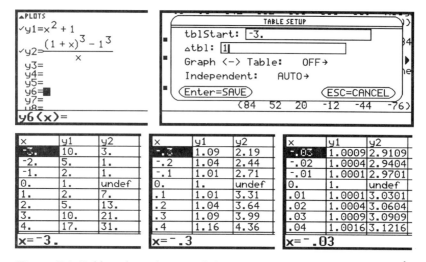

Figure 7.4 Table values closer and closer to zero, indicating that the y1 *values approach 1 and the* y2 *values approach 3.*

screen) to do a kind of zoom in close to zero. We get closer and closer by successively using `ΔTbl=1`, `.1`, `.01`, etc. At zero the second function is undefined, as shown in the table.

An unreliable table

Remember that by setting `INDEPENDENT` to `ASK` in the `TblSet` window, you can specify table values. The table in Figure 7.5 is more succinct than the previous ones. The new, third function is $j(x) = \sin(2\pi/x)$. The first seven lines of the table show even more persuasively that f approaches 1 and g approaches 3 as x approaches 0. The same reasoning suggests that

$$\lim_{x \to 0} j(x) = 0$$

PLOTS
$\checkmark y1 = x^2 + 1$
$\checkmark y2 = \frac{(1+x)^3 - 1^3}{x}$
$\checkmark y3 = \sin\left(\frac{2\cdot\pi}{x}\right)$

x	y1	y2	y3
1.	2.	7.	0.
.1	1.01	3.31	0.
.01	1.0001	3.0301	0.
.001	1.	3.003	0.
.0001	1.	3.0003	0.
.00001	1.	3.	0.
1.E-6	1.	3.	0.
.003	1.	3.009	.86603

x=.003

Figure 7.5 Selected values may be unreliable for guessing the limit.

The numeric data has led us astray. This function is similar to the one used for the 'inaccurate graph' in Chapter 4; its graph hints that it has no limit. A value that is not the reciprocal of an integer, such as 0.003, shows that the function is not going to zero. The limit of $j(x)$ as x approaches 0 does not exist.

Speeding ticket: the Math Police let you off with a warning

A warning: The graphic and numeric evidence may be very strong to indicate what the limit value should be, but don't make a speedy decision, as this can lead to the wrong conclusion. Only by using careful mathematical analysis can you really prove that the limit exists. For example, we rewrite

$$g(x) = \frac{(1+x)^3 - (1)^3}{x} = \frac{(1+3x+3x^2+x^3)-1}{x} = 3 + 3x + x^2$$

and see that

$$\lim_{x\to 0} g(x) = \lim_{x \to 0} (3 + 3x + x^2) = 3$$

Other techniques are necessary for trigonometric functions.

The `Limit` function

The TI-92 can do some of the careful mathematical analysis to determine limits. The `F3` (`Calc`) pull-down menu includes `3:Limit`. Although entered in a linear fashion, limit commands are displayed in traditional mathematical format in the history area. In Figure 7.6 we directly find the three limits of the previous example.

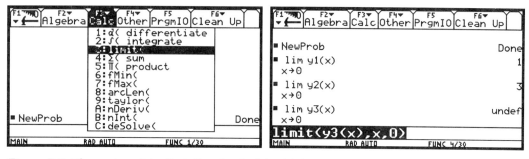

Figure 7.6 The syntax of the limit function is `limit` *(expression, variable, point).*

Local and long-term behavior

We can use the limit to tell us about local behavior of functions. For example, one classic difference equation can be simplified as

$$f(h) = \frac{\sin(0+h) - \sin(0)}{h} = \frac{\sin(h)}{h}$$

and we are interested in the local behavior close to zero. In Figure 7.7 we see that the limit is 1.

We look at two other rational functions, one where the limit is undefined and one where the limit is positive infinity. Both of these functions have a vertical asymptote at $x = 2$. We will say more about the undefined limit below.

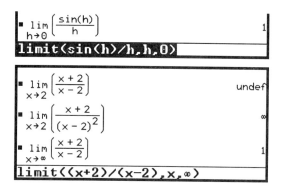

Figure 7.7 Finding local and long-run behavior for a function.

We can also consider the long-run behavior of a function by entering an infinite upper or lower limit. The last computation in Figure 7.7 shows that the rational function considered has a horizontal asymptote at $y = 1$.

Verifying the derivative

In the next chapter we formally define the derivative as the limit of a difference equation. We preview this in Figure 7.8 with three examples where a is an undefined variable.

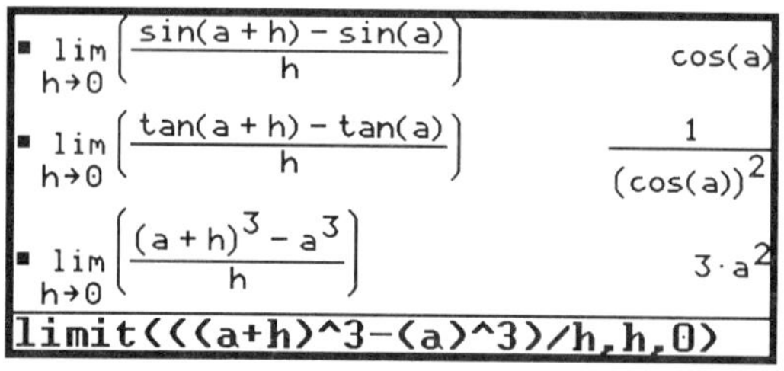

Figure 7.8 A preview of the use of a limit in finding a derivative.

Undefined results

When you are given the result `undef`, you must take care in interpreting it as actually undefined. The TI-92 uses several techniques, analytic and numeric, to do determine limits, but it is not foolproof. The response `undef` means that it did not determine a unique root, but one could exist, therefore it is better to interpreted this as "unknown." Figure 7.9 shows a case is shown where the limit is not given until a variable constraint has been made.

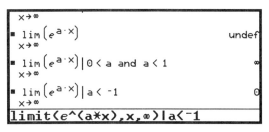

Figure 7.9 *The limit is not known until a variable constraint has been made.*

One-sided limits (optional)

There are some functions for which the limit as undefined at an *x*-value, but there is finite limit if we approach from only one side. The classic example is a step function, where the value jumps at given points. Such a function is $y = \text{int}(x)$. The graph and limits are shown in Figure 7.10. By specifying a direction as the fourth entry in the `limit` function, it finds the requested one-sided limit (notice the superscript + or - in the pretty print format). Because the limits from the left and right do not agree, the limit as *x* approaches 1 (from both sides) does not exist. We also show

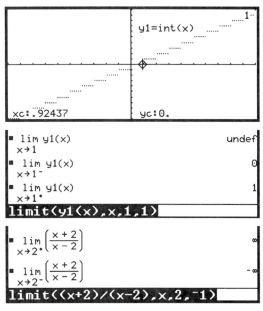

Figure 7.10 *One-sided limits. Because of the screen scale the superscript + is hard to read.*

that one of the rational functions from Figure 7.8 has limit $-\infty$ as *x* approaches 2 from the left and limit $+\infty$ as *x* approaches 2 from the right. Remember that one-sided limits can also be undefined; the limit of $\sin(2\pi/x)$ is undefined as you approach $x = 0$ from the right or left.

➤ *Tip: The direction value is read only as positive or negative — the magnitude is ignored. For example, to compute the right-hand limit as x approaches 5, you enter 1 (positive) as a direction value even though x = 1 is to the left of x = 5. It is best to use only 1 and -1 as the direction entries when computing a one-sided limit.*

8. FINDING THE DERIVATIVE AT A POINT

The derivative at a point is the instantaneous rate of change that we mentioned in the previous chapter. The average rate of change in miles per hour for a trip in your car is calculated by dividing the distance by the time. When you glance down to look at your speedometer, you are looking at the rate of change of distance with respect to time at that instant. In this chapter we will look at a graphical interpretation of the derivative at a point and learn three ways to use the TI to calculate this value.

Comparing derivatives at a point: three methods

In Chapter 4, we saw that we could set a center point and zoom in for a microscopic view of a graph. With repeated zooming, the graph began to appear linear. If we can zoom in until the graph appears linear, then we are essentially seeing the line tangent to the graph. The slope of this tangent line is the derivative at the center

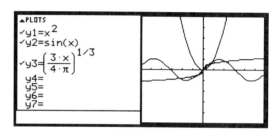

Figure 8.1 Three classic functions for investigating the derivative. (Split screen.)

point. More efficiently, you can use the F5 (Math) menu to draw a tangent line to a graph at some point, as detailed below.

We will consider three classic functions in this chapter:

$$f(x) = x^2, \quad g(x) = \sin(x) \quad \text{and} \quad r = h(V) = \left(\frac{3V}{4\pi}\right)^{\frac{1}{3}}.$$

The last function is the radius of a sphere in terms of its volume. The three functions have been graphed in Figure 8.1 using a split window and the ZoomDec window setting. Can you see which function is which? (If your sine function is flat against the *x*-axis, then you need to set MODE Angle to RADIAN.)

At the point $x = 1$, which function has the greatest rate of change? We could look at the graph, perhaps zooming in, and compare the steepness of the three functions at $x = 1$. To make a more exact comparison, we will evaluate the derivative of each function at $x = 1$. These three functions allow us to explore three different ways the TI can perform that computation.

Slope line as the derivative at a point: TANLN

Let's look at the parabola, y1, first. Turn the other functions off (recall this is done on the Y= editor screen with F4 (✓)). To draw a tangent line, from the

8. FINDING THE DERIVATIVE AT A POINT

GRAPH window use F5 (Math) A:Tangent. You are prompted to select the point where you want a tangent line to be drawn: move the cursor to the point or enter its x-value and then press ENTER. See Figure 8.2.

The tangent line is drawn and its equation is given at the bottom of the screen. We see that the slope of the line tangent to the parabola at (1, 1) is 2, so this is the derivative of the function at $x = 1$.

➤ *Tip: Should you want to draw a second tangent line, realize that the first one will remain on the screen unless you use F4 (Regraph) to start with a fresh graph.*

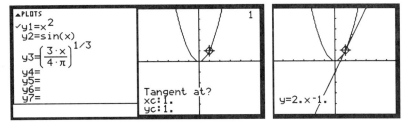

Figure 8.2 Using the F5 (Math) submenu selection A:Tangent to find a tangent line.

The derivative at a point, without the tangent line: dy/dx

Tangent lines are like training wheels on a bicycle: they eventually become unnecessary. We now show how to just find the derivative at a point. In Leibniz notation, the symbol for the derivative is *dy/dx*. This is the first selection of the 6:Derivatives▸ submenu available under F5 (Math) in the GRAPH window. Figure 8.3 shows finding the derivative of the sine function at $x = 1$. This derivative value, near one-half, is less than the derivative of $f(x) = x^2$ at the same x-value; in comparison, the sine values are increasing more slowly $x = 1$. Our next example shows a non-graphical approach to finding the derivative.

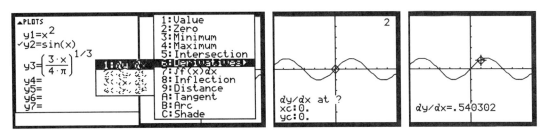

Figure 8.3 Using F5 (Math) 6:Derivative 1:dy/dx to find the derivative at a point on the graph.

▶ *Tip: Choices 2, 3 and 4 in the derivative submenu are available only when graphing in the 3D mode. Even though the submenu appears to the left, you press the right arrow to activate it.*

Using a symbolic approach to finding the derivative at a point: d(...)|x=a

We show with the remaining function a non-graphical approach to finding the derivative. From the HOME screen we can find the derivative value of y3 by using the F3 (Calc) menu option 1: d or 2nd_d (above the 8 key) and typing

d(y3(x),x)|x=1

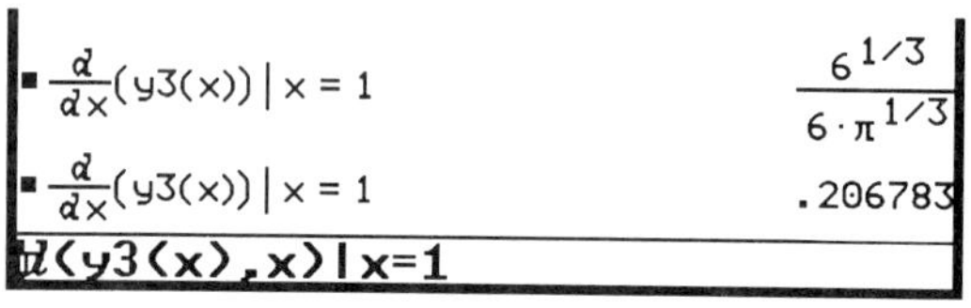

Both the selection from the menu and the keyboard option give the italicized letter and a left parenthesis. The bar 2nd_| is above the K key and may be read as "such that" or "with." This adds a condition to a command; entering d(y3(x),x) would give the derivative function, the topic of the next chapter. This full command above has the TI compute the derivative function of y3 and then evaluate it at $x = 1$.

Figure 8.4 Using d to find the derivative of a function at a point.

Figure 8.4 seems to give two responses to the same entry, but we pressed ◆_≈ the second time to give a decimal approximation of the answer. The TI-92 output differs in a minor way from the exact answer shown here; it is in a less succinct but equivalent form (the decimal approximations are the same).

The basic calculus commands are in the menu F3 (Calc). They don't look like any standard calculus notation when you type them in, but they are translated to traditional form in the history area. In the last chapter, we used 3: limit from this menu. The command nDeriv we will be explained very soon, the integral calculus commands ∫ and nInt will be used heavily in Part III of this book, arcLen in Chapter 18, and taylor in Chapter 19.

In our use of the derivative command, the first entry in d(...) is the function, the second entry is the independent variable of the function, and the "such that" bar precedes the identification of the evaluation point. The function need not be in the Y= editor and therefore x does not have to be the variable: d(t²,t)|t=1 gives a value of 2, just as we found for y1 above. The variable assignment after the "such that" bar applies only for this one command, in contrast to the more lasting effect of the entry 1→t. Most important, remember you must always identify the independent variable: you cannot enter y1'(1) to evaluate the derivative (to dissuade you, the single quote is not identified in yellow).

In summary, we have found that $f'(1) = 2$, $g'(1) \approx .5$, and $h'(1) \approx .2$, so the functions are given in descending order of their rates of change at $x = 1$.

Comparing the exact and the numeric derivative

The `F3` (`Calc`) menu offers another derivative command, `nDeriv` for 'numerical derivative.' Why two options? There are advantages and disadvantages to each.

Limitations of d

The symbolic derivative command is limited because it finds values using a formula-based approach, like knowing that $y' = \cos x$ for $y = \sin x$. In the absence of a list of known functions, you should be prepared for an error message if your function is out of the ordinary. For example, the function `int` is not on the list, so the expression `d(int(x),x)|x=.5` comes back unevaluated (all it does is replace `int` with the technical name `floor`). In such cases, we can use the numeric derivative command.

This numeric derivative is not limited by the calculator's compendium of symbolic derivatives and will find values for any function. For instance, we see in Figure 8.5 that this command gives an answer for the derivative of $y = \text{int}(x)$ at $x = .5$ that `d` did not recognize. But as with almost all numeric approximation algorithms, `nDeriv` sometimes gives false results.

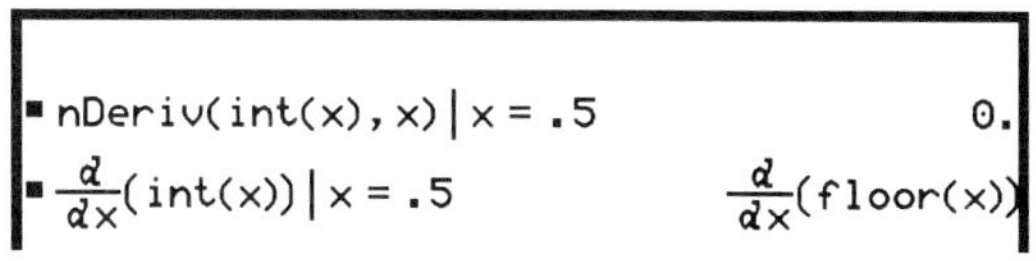

Fig 8.5 Symbolic and numeric attempts at differentiating the greatest integer function.

False results from nDeriv

The fact that `nDeriv` is an approximation can get us into deep trouble with certain points of some functions. For example, we know the function $f(x) = 1/x$ is not defined at zero and thus has no derivative there. We see in Figure 8.6 that using `d` to find the derivative at $x = 0$ gives $-\infty$, i.e., the derivative does not exist. But `nDeriv` erroneously gives a value of 1 million. (Neither of the graphic derivative methods described earlier will give an answer from the graph of $y = 1/x$ when we enter $x = 0$.)

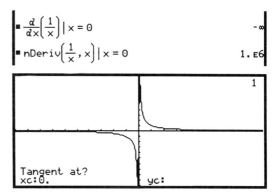

Figure 8.6 The home screen derivatives can give false results. From a graph no derivative or tangent will be allowed at an undefined.

This problem with nDeriv is not restricted to the obvious cases where the function itself is undefined. Our function y3=(3x/(4π))^(1/3) is defined for all real values, so you might expect to find a derivative at $x = 0$ in the same way we did for $x = 1$. However, the tangent line at $x = 0$ is vertical. This means that the derivative is undefined there (since a vertical line has no slope). Figure 8.7 shows this situation where, again, nDeriv gives a wrong answer and d responds with undef. (The two graphical derivatives give No solution error messages when we enter $x = 0$ from the graph of y3.)

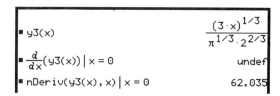

Figure 8.7 Invalid responses from nDeriv where d gives the correct answer.

Choosing between d and nDeriv

At worst, d is limited: it never gives an incorrect answer, but may respond with undef even though the derivative can be computed. When your function is beyond the TI's ken, use nDeriv — with caution. The last section of the chapter gives details of nDeriv and how you can improve its output.

There is another issue to be aware of with both derivative options. Since the domain of the logarithm is positive numbers, $\ln(x)$ should not have a derivative for negative x, but the top of Figure 8.8 shows that both commands give a value for the derivative of $\ln(x)$ at $x = -2$. The "problem" here actually has nothing to do with derivatives. Rather, the TI considers the logarithm in a broader context. We mentioned parenthetically in Chapter 6 that the calculator allows logarithms of complex numbers; the domain for this extended function is all non-zero complex numbers, including negative real numbers. The derivative of this logarithm is the same as the derivative of the more familiar real-valued function, thus the unexpected answers. Again, you would need a class in complex analysis to fully understand this. What you should remember is that the calculator may give familiar functions a larger domain than you expect.

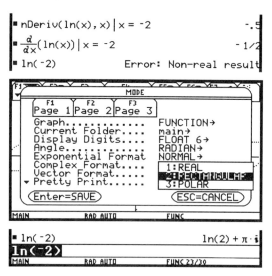

Fig 8.8 Logarithms of negative numbers and the MODE Complex Format settings.

How nDeriv works and how to improve its accuracy (optional)

You know from your calculus text that the derivative of f at a is defined as

$$\lim_{h \to 0} \frac{f(a+h) - f(a)}{h}, \text{ if the limit exists.}$$

(This definition can be created on the TI-92 by using the average rate of change function, avgRC, found in the CATALOG). The nDeriv function forms an altered "two-sided" form of the difference quotient:

$$\text{nDeriv(f(x),x,h)} = \frac{f(a+h) - f(a-h)}{2h}, h = .001$$

The two definitions yield the same result in the limit as h goes to zero.

The default value of $h = .001$ in nDeriv can be changed to increase the accuracy of our nDeriv approximation. Ironically, though, as we try to increase the accuracy by making h extremely small, we run afoul of the calculator's computational restrictions and our accuracy declines.

In Figure 8.9 we calculate nDeriv(2^x,x,h)|x=0 for increasingly smaller values of h. Of the h values shown, $h = .00001$ is closest to the true value of the derivative of $y = 2^x$ at $x = 0$. By using calculus or d, we know the true value of the derivative is ln(2) which to nine decimal places is .693147180. (View the full decimal values to verify that the output for $h = .00001$ is closer to ln(2) than the output for $h = .001$, even though they appear the same with the standard setting that displays six decimal places.)

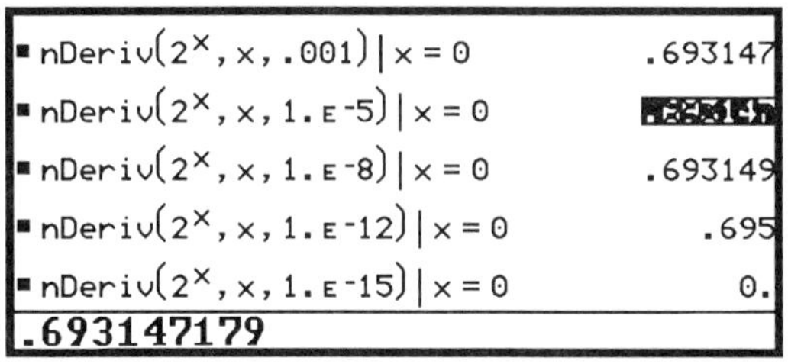

Figure 8.9 Resetting the h parameter to increase accuracy and going too far.

➤ *Tip: Use 2nd_EE to enter small values. Use negation, not subtraction.*

9. THE DERIVATIVE AS A FUNCTION

If we know the value of the derivative at a whole set of points, then we can define a new function called the derivative of f:

$$f'(x) = \lim_{h \to 0} \frac{f(x+h) - f(x)}{h}, \text{ if the limit exists.}$$

The computer algebra system of the TI-92 can often give us the derivative function in symbolic notation through the command **d**. For the comparably few functions beyond its recognition, we can create a function with **nDeriv** (numeric derivative) that is correct "most of the time."

But even with this strong symbolic capability, we believe it is crucial to explore the graphs of a function and its derivative. In addition to finding the formula for the derivative, you need to understand the relationship between a function and its derivative.

Comparing the numeric to the symbolic derivative

The derivative at a point was defined in the last chapter and we showed that the **F3** (**Calc**) menu commands

$$d(x^2, x) \text{ and } nDeriv(x^2, x)$$

could be evaluated at a point by using the "such that" bar. This means that they are functions themselves. The two commands for the squaring function give the same result, as shown in Figure 9.1.

Recall that **d** works from a list of known derivatives and differentiation rules while **nDeriv** uses numeric methods based on a version of the limit definition of the derivative. The **2.** in the **nDeriv** output for $y = x^2$ is a subtle indicator that it has arrived at this answer through computations. This is much more apparent when comparing, in Figure 9.2, how the two commands handle $y = \sin(x)$. This output from **nDeriv** is not useful as a symbolic answer, but it gives the correct answer when the derivative is evaluated at a point.

Figure 9.1 Symbolic derivative functions.

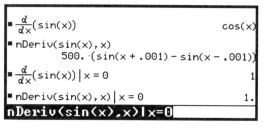

Figure 9.2 Contrasting **d** and **nDeriv** for symbolic and numeric derivatives.

As in the previous chapter, the conclusion is to use **d** whenever possible. As another example of a function that **d** does not know, consider **fPart(x)** which gives the fractional part of x, i.e., it ignores anything to the left of the decimal (except a negative sign). Figure 9.3 shows a graph of this function. By looking at the graph you can see that the derivative has a value of 1, except where it is undefined at non-zero integer values. The **d** function returns no formula and cannot be evaluated at a point.

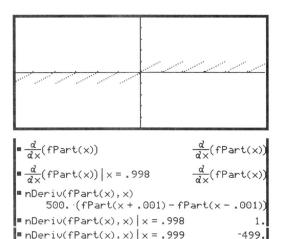

Figure 9.3 *Graph (with style* **2:Dot***) and derivatives of a function* **d** *does not recognize.*

The formula given by **nDeriv** is valid except within 1/1000 of non-zero integers. Decreasing h (as discussed in the previous chapter) will shrink these intervals where incorrect answers are given, but we can never get the correct answer of undefined for non-zero integers.

Viewing a graph of derivative function

Next we look at the graphical meaning of the derivative.

Matching a function to the graph of its derivative

We consider again the three classic functions from the last chapter:

$$f(x) = x^2, \quad g(x) = \sin(x) \quad \text{and} \quad r = h(V) = \left(\frac{3V}{4\pi}\right)^{\frac{1}{3}}.$$

The three functions were stored in **y1**, **y2**, and **y3** and graphed using a **ZoomDec** window setting. Now we graph the derivative function for each of the three functions and display them in Figure 9.4. Can you match the curves to the derivative functions?

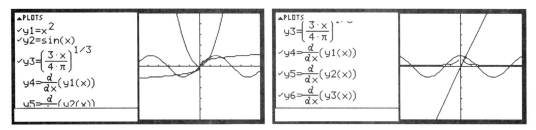

Figure 9.4 *Three derivative functions for three classic functions.*

We could, of course, graph them one by one, but using a little thought we can identify them just by looking at the features of the graph. Consider the parabolic function *f*: it is decreasing until it reaches zero, then it is increasing. Since the derivative gives the instantaneous rate of change, this means that the derivative values are all negative to the left of the origin and are all positive to the right of the origin. Of the three options, this describes the line. (You may also know the power rule of derivatives that says the derivative of a quadratic function is a linear function.) Now consider the sine function: it oscillates between decreasing and increasing, so the derivative should oscillate between negative and positive. There is only one function that does this and it looks like a cosine function (another rule you may know). The remaining derivative function is always positive and has a spike at zero; this fits the slope patterns of the y3 graph. As a reminder, the derivative function of *h(x)* is undefined at zero, but nDeriv erroneously gives a value there.

Graphing a function and its derivative in the same window

It is instructive to graph a function and its derivative in the same window, but you may want a means of distinguishing which is which. This is accomplished by using the F6 (Style) menu in the Y= editor. By choosing 4:Thick, you can change the graphing mode a bold line. Unfortunately, there is no indicator of a function's style save pulling down the Style menu for each function and looking for the option with the check to the left of its name. See Figure 9.5.

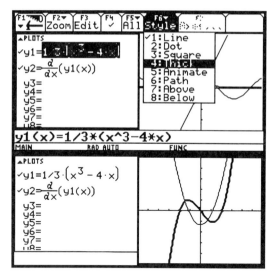

Figure 9.5 Two styles to distinguish a function and its derivative function in the same window.

Functions and their derivatives do not always fit very well in the same window. Consider a slight modification to the example above: add 100 to the y1 function, shown in Figure 9.6. The function graph will not appear in the ZoomDec window, but the derivative function will be identical to the derivative function shown in Figure 9.5.

There is a lesson here. Showing the graph and its function in the same window is a parlor trick and must be carefully designed to work. Further, it should be realized that a function serving as a model and its derivative will use different interpretations of the *y*-axis. For example, when modeling motion, the distance function might be in the feet and the derivative function would be in feet per second. The use of units is discussed in Part III Demonstration B.

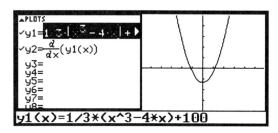

Figure 9.6 The function graph and derivative graph rarely fit in the same window.

The function that is its own derivative: $y = e^x$

Zero is an important number in addition because adding zero to a number does not change it. Similarly, one is important because multiplying a number by one does not change it. Could there be some function that is not changed by taking the derivative? In other words, could some function be its own derivative? To save time in guessing, we will try an exponential function. In Figure 9.7, we first change the function y1 to be our old doubling function $y = 2^x$ and graph it along with its derivative (leave y2 and deselect or clear y3). However, we will use our new style feature to have a leading circular cursor. This helps visually distinguish between two functions that have close values.

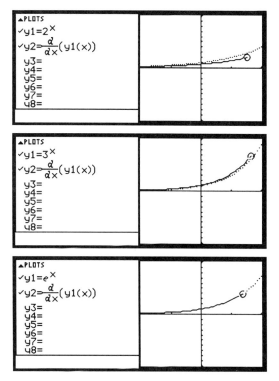

Now change y1 to the tripling function $y = 3^x$ and graph again. This function is very close to its derivative. The exponential function we are looking for has a base between 2 and 3. In calculus, we find that this amazing number is 2.718..., an irrational number denoted by *e*.

Figure 9.7 Looking for a function that is its own derivative.

The symbolic derivative of common functions

This brings us to the point where we trumpet the symbolic power to calculate derivatives. From the home screen you can request, and almost always get, the symbolic form of a derivative function. In Figure 9.8 we show the symbolic derivatives for several common functions. The history area looks like the list that appears in the back cover of many popular calculus books. Although it is reassuring to know they can be pulled up quickly, you should know these formulas by heart.

➤ *Tip: The TI-92 without Plus gives some slightly different algebraic forms.*

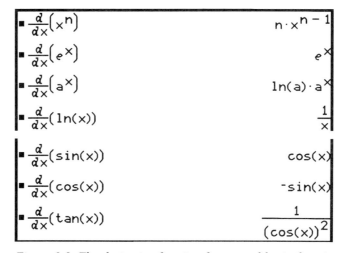

Figure 9.8 *The derivative function for several basic functions.*

10. THE SECOND DERIVATIVE: THE DERIVATIVE OF THE DERIVATIVE

In Chapter 8 we defined the derivative of a function, and in Chapter 9 we saw that the derivative itself is a function. Considering the derivative as a function, there is nothing stopping us from finding *the derivative of the derivative*. This is called the second derivative of the original function $f(x)$, written $f''(x)$. The TI-92 differentiation command d has a third optional input that allows us to take second (and higher) derivatives. The full syntax is

d (function, variable [, order])

and the default setting for *order* is one. So enter d(f(x),x) for $f'(x)$ and d(f(x),x,2) for $f''(x)$.

In the first section of this chapter, we show the second derivatives of several common functions and begin establishing the graphic relationships between the function and its derivatives. Then we consider the logistic function, a frequent model for population growth, using its second derivative to discuss the concavity of its graph. Finally, we revisit the grapefruit from Chapter 7 and examine an application of the second derivative to physics.

Symbolic derivatives and their graphs

We begin with the second derivatives of three well-known functions. The power rule answer will differ on TI-92s with and without the Plus chip. The actual symbolic format given by any symbolic algebra system depends on the internal rules of substitution and simplification. Often the algebraic complexity of the derivative increases with the order of the derivative. This is caused by the differentiation rules discussed in the next chapter.

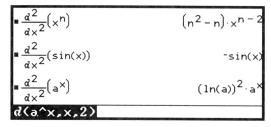

Figure 10.1 Second derivatives of common functions.

Now consider the simple function $y = x^2$. We enter it as y1 and define y2 and y3 to be its first and second derivatives. On the left side of Figure 10.2, we define the

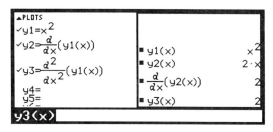

Figure 10.2 The first and second derivatives of $y = x^2$ symbolically.

functions and on the right (HOME) screen we can see the symbolic derivative functions, along with a verification that $(y')' = y''$.

Let's look at this graphically. In Figure 10.3, the GRAPH screen (using ZoomDec) shows the three graphs, using Thick for y1, Line (the default) for y2, and Dot for y3. Recall our discussion from the last chapter about how the line through the origin (y2) gives the rate of change of the parabola (y1). Considering y3 as the derivative of y2, the rate of change is simply the slope of y2, so y3 is the horizontal line $y = 2$. Considering y3 as the second derivative of y1, it measures the concavity; $y = x^2$ is the curve that has a constant positive concavity of 2.

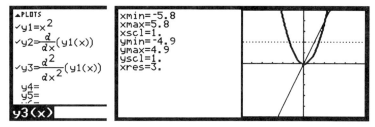

Figure 10.3 The first and second derivatives of $y = x^2$ graphically.

Graphs using d (and even more so nDeriv) in their definitions require calculations that slow the process considerably. By setting xres=3, only one-third of the usual x-values will be evaluated, thereby increasing the speed. Another benefit from this is that the dot style of the second derivative graph will be more pronounced.

▶ *Tip: If the graph is taking a very long time, press ON to stop it and reset xres to improve the speed.*

We have commented that, in general, trying to graph a function and its derivative in the same window is not practical. This is even more of a challenge with the first and second derivatives. We see this in the next example as we analyze the logistic function and its derivatives.

Looking at the concavity of the logistic curve

In most growth situations, a logistic growth model makes more sense that a pure exponential model. A new software company may keep doubling employees, but the growth has to slow down or else, like the paper folding that reached the moon, the number of employees would eventually be out of this world, literally. Suppose the logistic function entered as y1 in Figure 10.4 gives the number of

10. THE SECOND DERIVATIVE: THE DERIVATIVE OF THE DERIVATIVE

employees in the company at time x. (There is no need to change y2 and y3; they will refer to this new y1.)

We want to view the first and second derivatives and see what they tell us about the logistic function. It is necessary to consider each one individually, finding an appropriate window for each graph. The graph y1 has a range of values that would make seeing y2 and y3 impossible in this window.

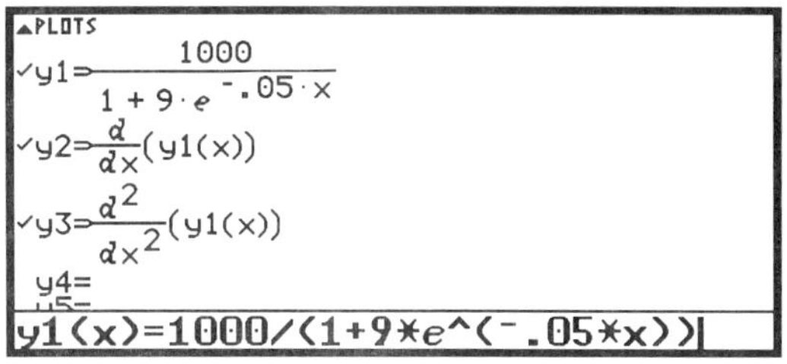

Figure 10.4 Defining a logistic function and its first and second derivatives.

The function y1 is monotonically increasing on the interval shown in Figure 10.5, so the derivative function y2 must be positive there. Notice that y2 peaks and then decreases, approaching zero. The peak is at the point of the fastest growth of y1. The second derivative shows the rate of change of the rate of change. We see that the value of y3 is zero at the peak of y2. (The domain for x is the same in each screen, but we change the range for y to display the behaviors of each function.)

In this example we can see that the second derivative (y3) is zero at about 40. This is because the growth rate (y2) peaks at about 40 and begins to slow. On the graph of y1, this is the point of fastest growth. It is a point where the concavity is changing from positive to negative. We call this a point of inflection on the graph. Since this is an important point on y1, we want to identify it more exactly.

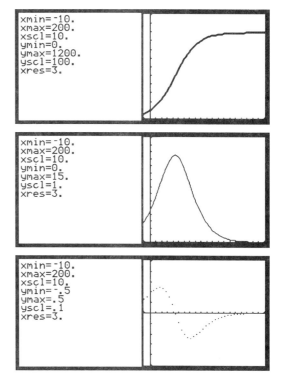

Figure 10.5 The logistic function and its two derivative functions.

Our way of identifying this point is to find the zero of the second derivative. Recall that Zero is an option on the F5 (Math) pull-down menu on the GRAPH screen. The computation is shown in Figure 10.6. As you reproduce these screens on your calculator, you will notice that y3 is drawn much more slowly than you are accustomed to. Even moving

76 PART II / DIFFERENTIAL CALCULUS

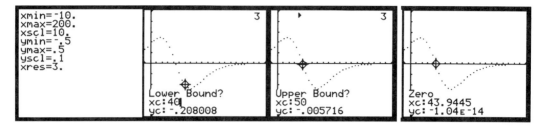

Figure 10.6 Using **2:Zero** *to find the zero of the second derivative (remember that the y-value may be only extremely close to zero).*

left and right with the cursor requires some patience. This is because y3 is not drawn from its own formula, rather from an operation on y1. Were we to put in the formula for y3, it would graph much more quickly.

These same computations can be done symbolically, at the expense of the graphical insight. The derivative formulas are not especially enlightening, but the computations are faster. Notice that the calculator "thinks of" y1 in a form other than what we entered. Figure 10.7 confirms the values we found above (using the F3 (Calc) option 7:fMax to find the highest value of y2).

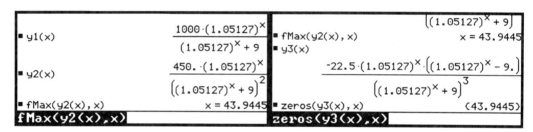

Figure 10.7 Symbolic computations of the point of inflection of y1.

Creating a numeric second derivative using a table

We now want to reexamine the data from the beginning of Chapter 7 about the height of a grapefruit tossed into the air over six seconds. Recall that we store variables and functions in long names so that, unless you erased them using 2nd_VAR-LINK, the information should still be there. If not, you can simply enter it now with the values shown in Figure 10.8 The delta function was defined as

Define delta(x)=seq(x[i+1]-x[i],i,1,dim(x)-1,1)

Recall the average velocity computation delta(height)/delta(time) was our best guess of the derivative for each of the first six seconds. The derivative of velocity (the second derivative of height) is acceleration. We can compute the acceleration from the original data by computing delta of the average velocities divided by 1, the time intervals. Notice that we do not have to

10. THE SECOND DERIVATIVE: THE DERIVATIVE OF THE DERIVATIVE

give the average velocity list a name as we can paste it in from the history area. Also, it does not work to divide by `delta(time)` since that list has seven entries and the velocity list only has six because of the way it was computed.

The list produced at the end of Figure 10.8 has five identical entries of -32. This number may be familiar to you as the downward acceleration due to gravity (measured in feet per second squared). Computing the second derivative from the tabular data of heights reminds us of the constant effect of gravity.

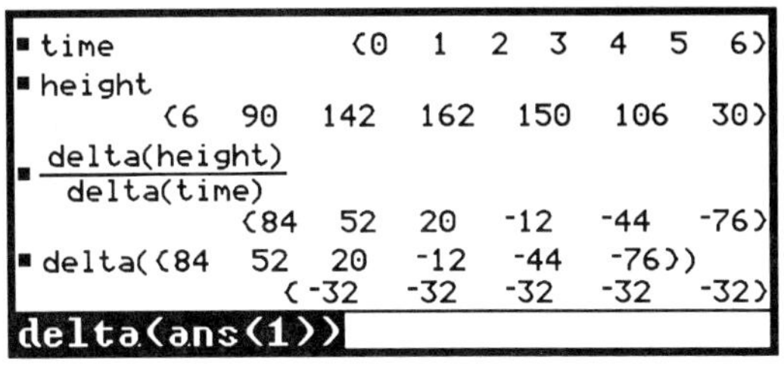

```
■ time                    (0   1    2    3    4    5    6)
■ height
              (6   90   142  162  150  106   30)
■ delta(height)
   ─────────
   delta(time)
              (84   52   20  -12  -44  -76)
■ delta((84   52   20  -12  -44  -76))
              (-32  -32  -32  -32  -32)
delta(ans(1))
```

Figure 10.8 Computing accelerations from a list of velocities.

11. THE RULES OF DIFFERENTIATION

Using the definition of the derivative to find a derivative function is cumbersome; fortunately there are shortcuts to finding derivative functions. We can use the symbolic feature of the calculator to show the common rules. These rules can be proved analytically, but we will do a graphical verification with two specific functions to encourage your intuition and strengthen graphic understanding. Also, we will show graphically that some common guesses for the rules are wrong.

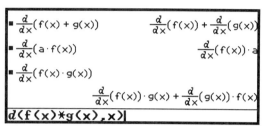

Figure 11.1 Several differentiation rules.

The Product Rule: (fg)' = f'g + fg'

We take two common functions and graph their product and the derivative of the product, as shown in Figure 11.2. We see from the graph that the derivative function is zero at $x = -2$, corresponding to the local maximum of the product function there.

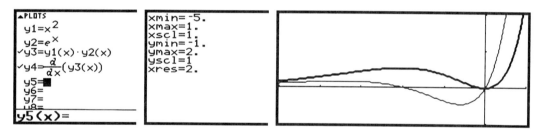

Figure 11.2 A product function and its derivative (the style for y3 is Thick).

Now let's suppose we make the tempting guess that the derivative of the product is the product of the derivatives. This is not unreasonable since the derivative of a sum is the sum of the derivatives, but it is wrong. We can see this

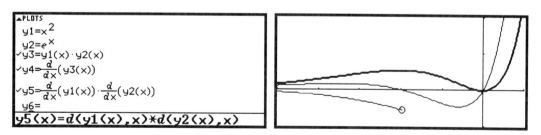

Figure 11.3 The product of the derivative functions being drawn with the Path style does not follow the derivative function nor does it have a zero at the product function maximum, therefore $(fg)' \neq f'g'$.

in Figure 11.3 by graphing the product of the derivatives. The product of the derivative functions does not have a zero where it should, at $x = -2$, so it cannot be the derivative of the product function.

To see that the product rule holds for these two functions in this window, we enter the true formula and use the `Path` style in Figure 11.4 to see that now the y5 graph traces over the y4 graph.

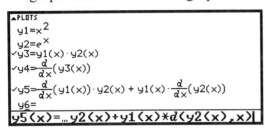 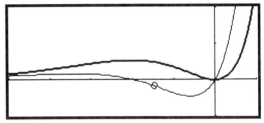

Figure 11.4 The corrected derivative formula function, y5, traces over y4, the derivative of the product.

Finally remember that the symbolic answer may be written in any number of equivalent ways and in Figure 11.5 the TI-92 Plus screen shows that the common factor e^x has been factored.

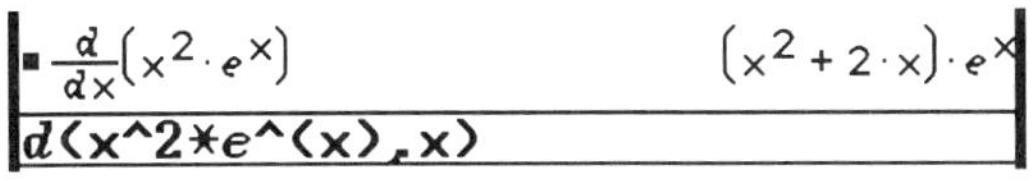

Figure 11.5 The product rule used to differentiate.

The Quotient Rule: (f/g)' = (f'g-fg')/g²

Let's check the quotient rule in the same way. First we generate the formula in Figure 11.6, but, as is common with computer algebra systems, the answer is not in the traditional form. The `F2` (`Algebra`) command `6:comDenom` gives the symbolic derivative as the familiar single ratio.

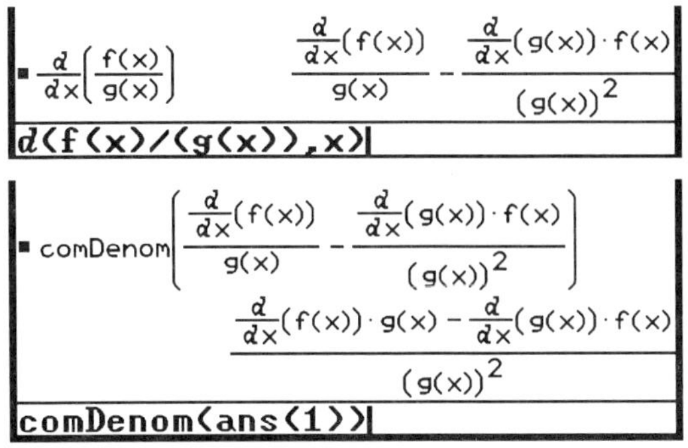

Figure 11.6 The quotient rule brought to familiar form by writing the symbolic derivative as a single ratio with the `comDenom` command.

Using our previous functions, we need only change y3 and make some minor adjustments to the window to get the graph shown in Figure 11.7. Notice that the two zeros of the derivative function are at the local maximum and minimum of the quotient function.

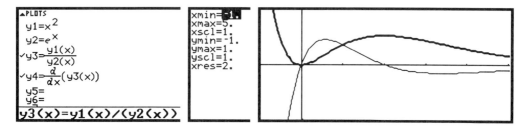

*Figure 11.7 A quotient function (**Thick**). Its derivative, y4, has zeros at the local extrema of y3.*

In Figure 11.8 we can again make a feasible but incorrect guess, namely that the derivative of the quotient is the quotient of the derivatives. We see our folly easily since the quotient of the derivatives is not zero at $x = 2$ (the local maximum of y3) and does trace over the known derivative function. Finally enter the correct formula in y5 (in the HOME screen, pasting from the history area and using STO▶) and see that it follows y4.

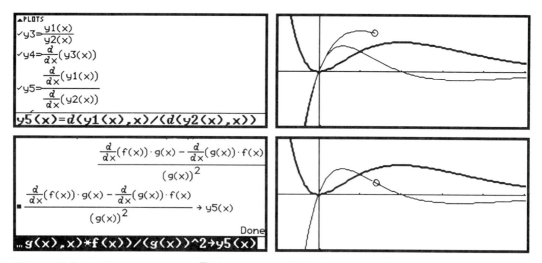

Figure 11.8 An incorrect guess, y5, does not match the derivative y4. The correct quotient rule formula does trace over y4.

The Chain Rule

Thinking of a function as a composite function and then using the chain rule will often simplify finding the derivative. Consider $y = (x^2+1)^{100}$. A straight-forward but impractical approach would be to expand the expression, write it as a polynomial of degree 200, and then differentiate term by term. Instead, we apply the chain rule and find the derivative quickly and easily:

$$y' = 100(x^2+1)^{99}(2x) = 200x(x^2+1)^{99}$$

Figure 11.9 shows an unsuccessful attempt to get the general chain rule formula using undefined functions f and g. However, once we define f and g, the chain rule is properly applied. We also verify the answer by finding the derivative of the composed function.

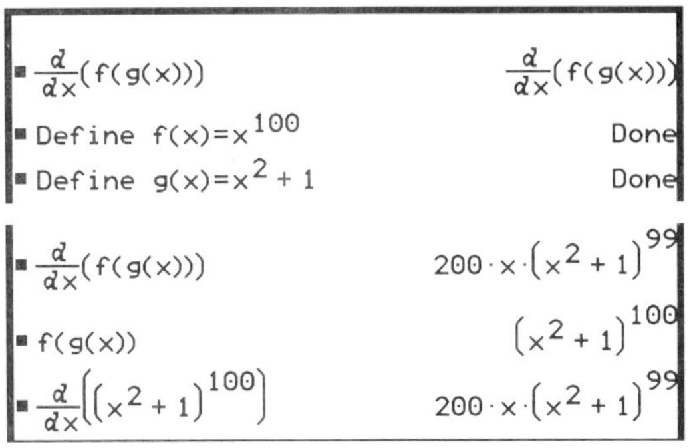

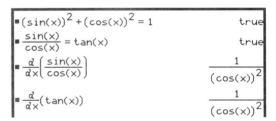

Figure 11.9 The chain rule cannot be shown in abstract form, but is used for specified functions.

The derivative of the tangent function

We can find the derivative of the tangent function by using the quotient formula on the definition $\tan(x) = \sin(x)/\cos(x)$. Applying the Pythagorean Theorem to the numerator of the derivative, the quotient rule gives $y' = 1/\cos^2(x)$. This is always positive and is undefined at multiples of $\pi/2$ (where the cosine is zero). In Figure 11.10 we verify the Pythagorean Theorem, check the abstract definition of the tangent, and verify that the derivative is the same for either expression. In a Figure 11.11 we compare these to the TI-92 without Plus to demonstrate two improvements made with the new chip. First, the TI-92 without Plus does not always handle equality of expressions with variables, so its output for the

Figure 11.10 TI-92 Plus screen concerning the derivative of the tangent function.

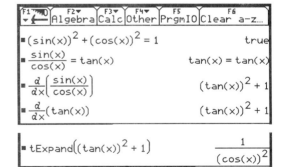

Figure 11.11 TI-92 without Plus screens to be compared with Figure 11.10.

second line is `tan(x)=tan(x)` instead of `true`. Second, the derivative is given in an unfamiliar form (the answer is the same for `d(sin(x)/cos(x),x)`). We use a trigonometric simplification to put the answer in simpler form. The command is under F2 (Algebra) in the 9:Trig submenu, 1:tExpand. Curiously, 2:tCollect gives the same output.

In Figure 11.12 we use the ZoomTrig window to view y1=tan(x) and y2=d(tan(x),x). This should remind you that *x*-values are sampled evenly across the window and the function values at these points are then connected to form the graph: the vertical lines should not be shown. (Using the Dot style would avoid this problem.) It should be clear from the graph that the tangent function is undefined at $\pi/2 + n\pi$ (for any integer *n*) and increases within each interval of definition such as $-\pi/2 < x < \pi/2$. Thus we expect the derivative to be undefined at the same values and positive on each such interval of definition. The second screen of Figure 11.12 confirms this. (There are no vertical lines here since the derivative function values sampled on either side of each $x = \pi/2 + n\pi$ are both large positive numbers).

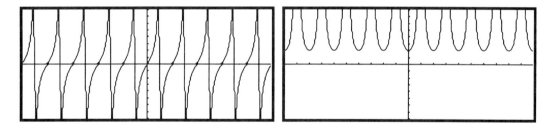

Figure 11.12 Graphs of `tan(x)` *and its derivative.*

➤ *Tip: The conventional mathematical way of writing a power of a trigonometric function, such as* $cos^2(x)$, *cannot be used on the TI-92. Instead, use* `cos(x)²` *or, better yet,* `(cos(x))²`.

12. OPTIMIZATION

One of the powerful uses of the derivative is helping maximize or minimize a function. But we must confess that calculators and computers with graphing capabilities can, in most cases, find maximum and minimum values of a function without your having to know anything about calculus. The examples of this chapter show both the calculus and the non-calculus approaches.

The ladder problem

Typically, optimization problems arise from real-world applications. The 'ladder' problem is to determine the longest ladder that can be carried horizontally around a corner that joins two hallways. We will assume that the hallways are different widths: the narrower one is 4 feet wide, the wider one 8 feet. Figure 12.1 shows the position where the ladder could get stuck: it touches both walls and the corner. For the

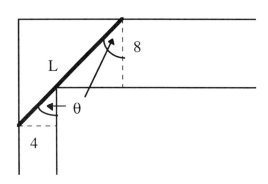

Figure 12.1 A ladder carried horizontally around a corner.

ladder to make this corner, it needs to fit in the hallway for every angle θ, even this tightest one. Think of the ladder as being divided at the corner point and use the triangle trigonometry definitions for the sine and cosine to find each piece. This leads to the following equation for the ladder's length in terms of the angle. We want to minimize

$$L = f(\theta) = \frac{4}{\sin(\theta)} + \frac{8}{\cos(\theta)}$$

A ladder of that minimal length will fit for all angles, and it will be the longest possible ladder that works since it touches both walls at the tightest angle.

Finding such an equation is the hard part of an optimization problem. We now show several different methods for finding the angle value that minimizes L. Before beginning, check the status bar for **RAD** to be certain that the angle mode is set to radians.

Graphic solutions without calculus

Enter L into ʏ1, using x as the variable. You might be tempted to use ZoomTrig setting, but, like most models, there is a more restricted domain in this example. The angle must be greater than 0 but less than $\pi/2$. To set the window y-values, we can be generous and say that the minimum ladder will be under 60 feet (we also set ymin=-10 to avoid any overlap with the numeric values at the bottom of the screen). Use Trace to approximate the lowest point on the graph, as shown in Figure 12.2.

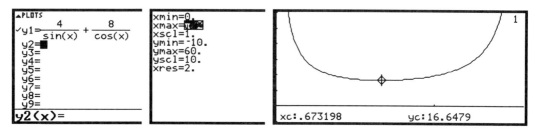

Figure 12.2 A minimum found by using Trace and the trial-and-error method.

Alternatively we can use F5 (Math) 3:Minimum to directly find $y \approx 16.6478$ (an improvement of one ten-thousandth). As shown in Figure 12.3, you must provide a lower bound and an upper bound. These values can be entered directly (the *x*-values must be within the viewing window) or by using the arrows and pressing ENTER.

We could also solve the problem graphically with calculus, but we defer that technique to the next example.

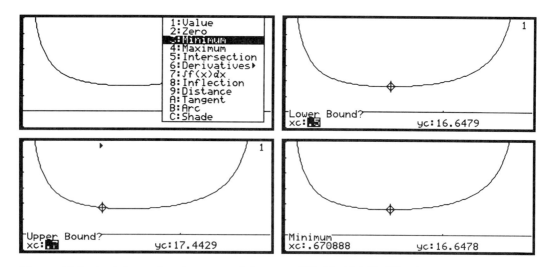

Figure 12.3 A minimum found on a graph by using F5 (Math) 3:Minimum.

Symbolic solutions

These same results can be found symbolically, both with and without calculus, from the HOME screen. The first screen of Figure 12.4 shows the F3 (Calc) option 6:fMin, first without constraints (which takes a few moments) and then with the conditions of the problem given after a "such that" bar (above the K key).

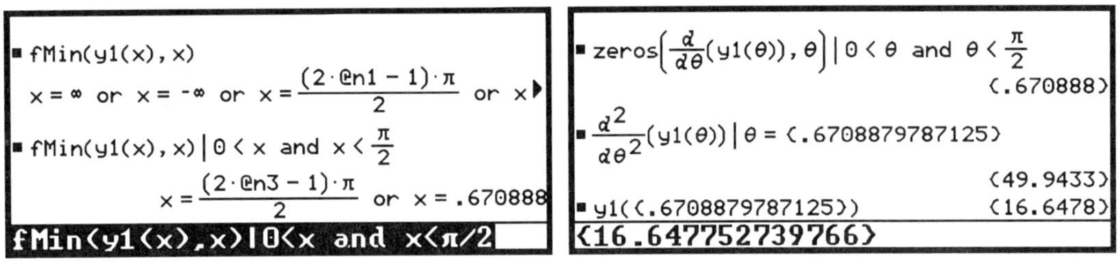

12. OPTIMIZATION

Figure 12.4 Two symbolic solutions with the final estimate of the minimum ladder length shown on the second entry line.

Notice that the interval constraints on *x* must be given as two separate inequalities.

The second screen of Figure 12.4 shows the calculus approach: find the zeroes of the first derivative in the restricted domain, apply the second derivative test to verify that the extremum is a minimum, and evaluate the critical value in the original function. Notice how we can use θ as a variable (the θ key is to the right of the M key).

Why bother with calculus?

After all of that, you may wonder why we bother to use derivatives. The first and most naive approach, using **TRACE** on the graph of y1, gave us an answer within 0.000147 of the most accurate solution. Perhaps there are situations where even a 0.0009% error cannot be tolerated. What we really hope for is an exact answer, which only derivatives can lead us to. Such 'closed form' solutions can be important for using in other calculations or for insuring unlimited accuracy.

The TI-92 did not give us an exact answer for this problem because it was unable to symbolically solve for the zero of the derivative. This is common particularly when trigonometric functions are involved. Figure 12.5 shows the TI derivative as a fraction. This will only be zero when the numerator is zero. We can solve for *x* by hand using algebra and find the exact answer.

In defense of the calculator, its symbolic reasoning makes no assumptions about the domain of values being $0 < x < \pi/2$, so it will not divide by $\cos(x)$, which could be zero.

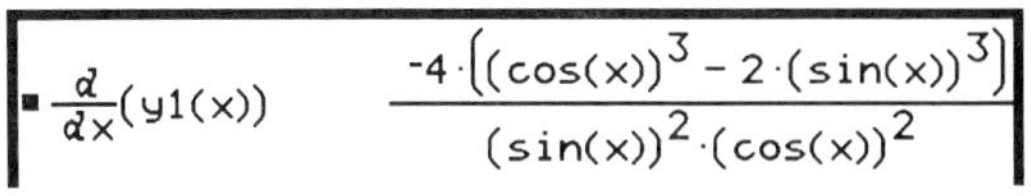

$$-4(\cos(x)^3 - 2\sin(x)^3) = 0$$

$$-4(1 - \frac{2\sin(x)^3}{\cos(x)^3}) = 0$$

$$\tan(x)^3 = \frac{1}{2}$$

$$x = \tan^{-1}(\sqrt[3]{\frac{1}{2}})$$

Figure 12.5 The derivative as a fraction and then solving for x by hand.

Recasting the problem

Another approach is to consider the problem in a different way. With the same reasoning as before, we see in Figure 12.6 that the shortest ladder touching the inside corner and the two walls is the hypotenuse of a right triangle with side lengths z and $y + 4$. By similar triangles, z is to $y + 4$ as 8 is to y. This allows us to write L as a function of y. Then we find the critical point, $y = 4 \cdot 2^{2/3}$. This is done in the left screen of Figure 12.6. In the right screen we classify the critical point as a minimum by using the first-derivative test, checking the sign of the derivative at the critical point ±0.1. The long-awaited closed form of the answer is shown. Finally, we force a decimal answer to see that it agrees with our previous answer in Figure 12.4 to the calculator's full twelve digits of accuracy.

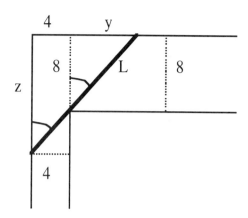

Figure 12.6 Another view of the ladder.

▶ *Tip: The second derivative test is commonly used in optimization problems, but in this case the second derivative found includes a* `sign` *function because of the radical involved. Since the* `sign` *values are constant, you can assume that its derivative is zero. With those terms zero, the second derivative evaluates to a positive number. (This calculation is not shown.)*

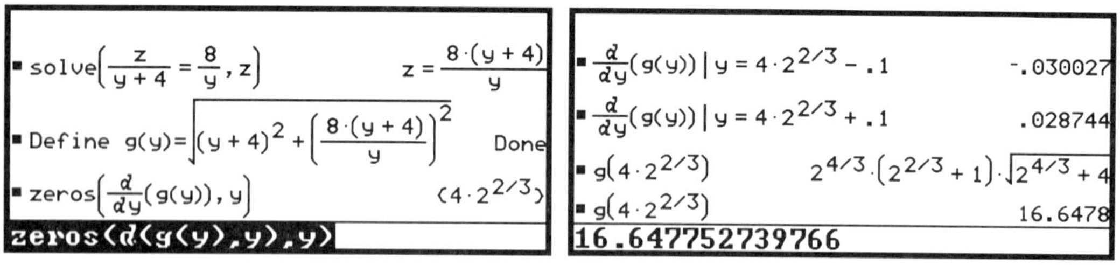

Figure 12.7 Another function giving the ladder's length and an exact form of its minimum value.

The first version of the problem follows the treatment in *Calculus* by Hughes-Hallett, et al., and the second version is based on Appendix 1 of the TI-92 *Guidebook*. Both are fully valid and give the equivalent results; one can show that

$$f(\tan^{-1}(2^{-1/3})) = g(2^{8/3})$$

The only difference is that the second version, having no trigonometric functions, is better suited for the TI-92's computer algebra system.

Box with lid

Suppose we have an 8.5 by 11 inch sheet of paper and want to cut squares and rectangles from the corners to create a folded box with lid, see Figure 12.8. We want to maximize the volume. Notice that if the x cut is very small, then the box will be so shallow that it hardly holds anything. If the x cut is large, then the bottom will be so small that the box again holds very little. We first find a volume function that depends only on the length x of the cut-out square:

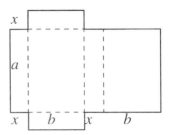

Figure 12.8 *A diagram of cutting corners to create a box with lid. Fold on the dotted lines.*

$$V = a \cdot b \cdot x = \left(\frac{17}{2} - 2x\right) \cdot \left(\frac{11 - 2x}{2}\right) \cdot x$$

We use 17/2 instead of 8.5 because having a decimal in the function definition would force all TI answers to be decimal approximations (although setting the **MODE** option **Exact/Approx** from **AUTO** to **EXACT** would override this). We will solve this symbolically, but first we take a graphical approach to see an approximate answer.

We want to graph the volume function along with its first and second derivatives. In Figure 12.9, we enter the function $V(x)$ as **y1**. To avoid slow graphing, compute the derivatives on the **HOME** screen and use copy-and-paste (♦_C & ♦_V) to put them into the **Y=** editor. The functions have been graphed using three different styles in the window shown. From the graph, press **F3** (**Trace**), move to the second function (down arrow), and find an approximate zero of the derivative. Notice that at the given x-value the volume function appears to be at a maximum and the second derivative is negative. This graphically confirms the calculus theory.

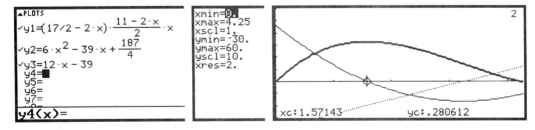

Figure 12.9 *Using **Trace** to find an approximate zero of the first derivative.*

88 PART II / DIFFERENTIAL CALCULUS

To find the maximum analytically from the HOME screen, we find the zeros of the first derivative. The multiple answers in Figure 12.10 can be treated as lists in subsequent computations. The second derivative test shows tell us the first critical point is a maximum and the second is a minimum. For simplicity we evaluate the volume function at both critical points (use y1(ans(2))). The exact answers can be shown in decimal approximation form with ◆_ ≈.

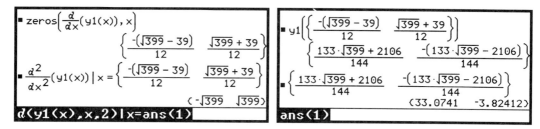

Figure 12.10 *Finding the maximum with the traditional calculus methods.*

➤ *Tip: Ideally, you will look at the graphs of the function and its derivatives before proceeding to the symbolic computations. If you choose to use only the analytic methods, be sure to use the derivative tests to classify critical values.*

Using the second derivative to find concavity

An important function in statistics is the *standard normal distribution* function. It has a graph that is bell-shaped and has the definition

$$f(x) = \frac{1}{\sqrt{2\pi}} e^{-x^2/2}$$

For simplicity we define a function in the same family with a leading coefficient of one. The equation and graph (Thick) are shown in Figure 12.11, along with the graphs of the first and second derivatives (their formulas were calculated and pasted in to the Y= editor).

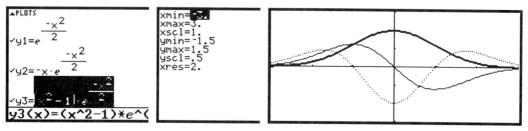

Figure 12.11 *A bell-shaped curve and its derivatives.*

We discussed in the last chapter that the second derivative measures the concavity of the graph. Points where the concavity changes sign are of particular interest; these are where the second derivative changes sign. Again, these are called points of inflection. Look at Figure 12.11. On the dotted curve, find the zero with positive *x*-value; the second derivative changes from negative to positive, so this is where the bell-shaped curve changes from being concave down to concave up. The other zero (with negative *x*-value) on the dotted curve is a point of inflection where the curve changes from being concave up to concave down. An important characteristic of this bell-shaped curve and its multiples (such as the standard normal distribution) is that they have points of inflection at $x = \pm 1$.

We confirm this in Figure 12.12 using two methods. We identify the point of inflection with negative *x*-value by using F5 (Math) 2:Zero to find the zero of the second derivative. Be careful to select y3 before entering the lower bound and upper bound. A more direct method is to use the option F5 (Math) 8:Inflection on y1 itself. This command has the same input sequence as Zero. By using Inflection, the derivative graphs are not necessary, but they are instructive since you can see that at the point of inflection, the first derivative has a relative extremum and the second derivative is zero.

There is no analytic inflection command that can be used from the HOME screen. Points of inflection would be *among* the list zeros(d(f(x),x,2),x). Just as every critical point is not a relative extremum, there are zeros of the second derivative where the concavity does not change. To determine which are points of inflection, you need to either consult the graph of the second derivative or check the concavity on either side of each zero (like the first derivative test but using the second derivative) — perhaps you should just look at the graph.

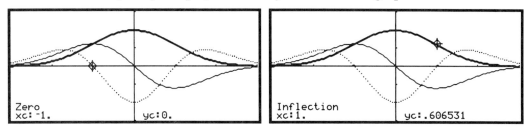

Figure 12.12 Using Zero *on* y3 *and* Inflection *on* y1 *to find points of inflection.*

A. OPTIMIZATION WITH CABRI GEOMETRY

This demonstration is designed to model an optimization problem using the Cabri geometry application of the TI-92. The geometry application has a rather steep learning curve and this will be an abbreviated presentation. You may need to first practice the geometry tutorials that are available in Chapter 7 of the *Guidebook*. The second part here shows how to use calculus to solve the same problem.

The bus stop problem

Suppose you are across a park from a bus stop and want to get there as quickly as possible. The bus stop is 2000 feet west and 600 feet north of the starting position. You can walk through the park terrain at a speed of 4 feet/sec and walk along the sidewalk at a rate of 6 feet/sec. What path will provide the fastest route to the bus stop? (This problem is outlined on page 272 of *Calculus*, Hughes-Hallett, et al.)

The geometry construction

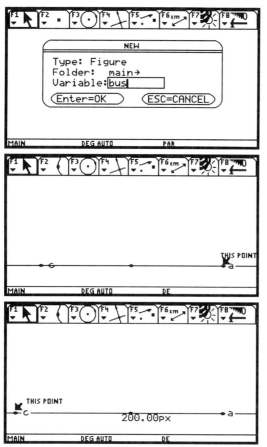

Use the key sequence APPS 8 3 to create a new geometry figure. We will name it **bus**. As it turns out, it is easier to scale down the distances by 10 and use Pixel units. Use ♦_F to change Length & Area to 1:PIXELS.

Using F2 4:Line, we first place a sidewalk line near the bottom of the screen and label two points (F2 1:Point) that will soon be 200 pixels apart.

Use F6 (cm) 1:Distance & Length to measure the distance between points a and c. Then hold down the hand-key while using the arrows to drag either a or c until the distance between them is exactly 200 pixels.

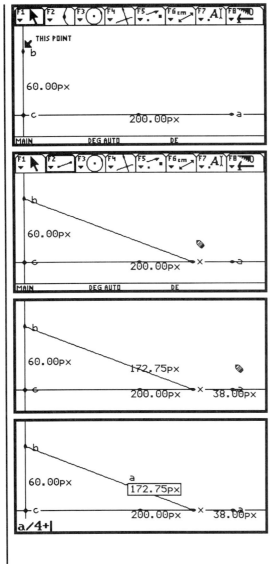

Construct a line perpendicular to the sidewalk line through the point c with F4 1:Perpendicular Line. We mark the bus stop as point b and use the hand to drag b to an exact distance of 60 pixels from c.

Now place a point x on the sidewalk where you will cut across the park. Use F2 5:Segement to connect the points x and b. This completes the geometric construction; we proceed to calculations now.

Use F6 (cm) twice to measure the distance between point a and x (shown as 38.00px) and then the length of the line segment from x to b (shown as 172.75px).

Enter the time formula with F6 (cm) 6:Calculate. The up arrow cycles through the four numbers on the screen. When you press ENTER on 172.75px, it will temporarily be labeled a (not to be confused with point a). Press ÷ and 4 (this gives the time across the park) and then + to add the second time part.

Finish the formula by arrowing to the 30.00px number and pressing ENTER. This becomes b which we divide by 6. Notice that the first part of the formula is temporarily erased.

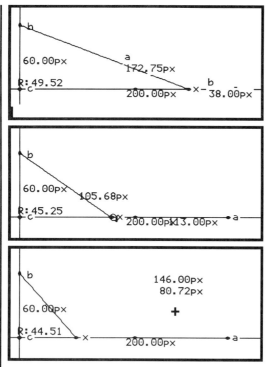

The calculated time is shown as R:49.52.

Use the hand to drag x along the sidewalk; notice that the total time value is decreasing. The distance from a to x crashes into the 200.00px measurement.

We drag the two path distances up away from the axis to make them more readable. Keep moving x to where time is a minimum. (Remember each number shown must be scaled up by ten to match the original context.)

Figure A.1 A geometric approach to optimization.

A calculus solution

The answer of 1460 feet along the sidewalk is only approximate. You may have noticed that 1470 feet also gives a time of 445.1 seconds. We now use calculus to give us exact answers.

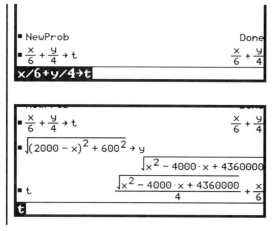

We start by using F6 (Clean Up) 2:NewProb to clear variables (on the TI-92 Plus). Let x be the distance on the sidewalk, y the distance through the park, and store the time formula as t.

By the Pythagorean theorem, we can write y in terms of x. This means that t is now defined completely as a function of the single variable x — the TI-92 automatically makes this substitution.

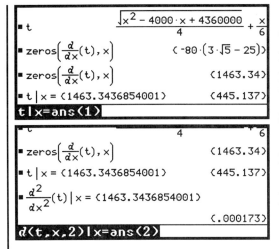

Now we apply calculus by finding the zero of the first derivative. The first answer is in exact form, the second in the more practical decimal form. Time is found by evaluating *t* with this *x*-value.

Finally, we verify that the critical point is at a minimum by checking to see that the second derivative is positive.

Figure A.2 A calculus-based numerical solution to the optimization problem.

B. IMPLICIT DIFFERENTIATION

Implicit differentiation is a common technique in calculus that requires algebraic dexterity. The TI-92 can do the algebra, but needs a little help in identifying relationships and renaming the derivative.

Let's take the simple example of the equation of a circle of radius a centered at the origin, $x^2 + y^2 = a^2$. If we just enter it that way, the calculator will assume that y is just another constant like a, so we must identify the relationship between y and x. In Figure B.1, we do this by writing $y(x)$ instead of y.

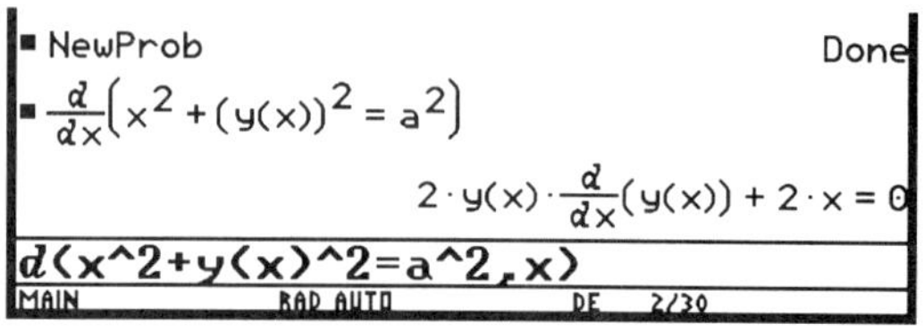

Figure B.1 Implicit differentiation of an equation.

Although it cannot be shown in Figure B.1, there will be a warning on the status line, `Warning: Differentiating an equation may produce a false equatic` (the intended last word, `equation`, doesn't quite fit).

The calculus is done, but the algebra remains. We want to solve for dy/dx. The TI-92 only allows you to solve for a variable, so we need an intermediate step where we rename `d(y(x),x)`. We choose an identifiable name, `yprime`. This renaming is easily accomplished by using `ans(1)` and specifying this condition after the "with" bar (`2nd_|`) as shown in Figure B.2.

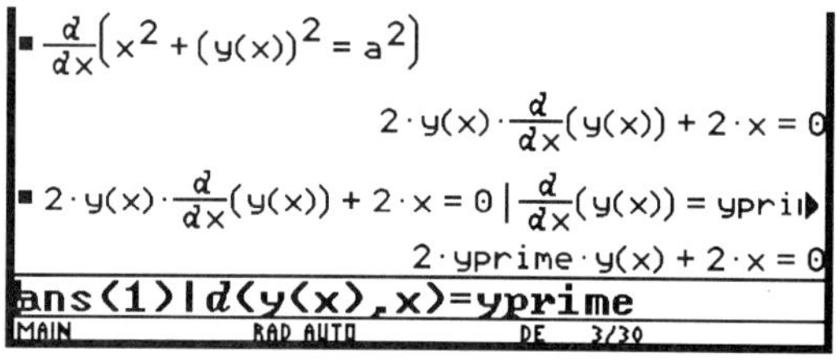

Figure B.2 Substitute a variable name for the derivative.

In the final step, shown in Figure B.3, we use `solve` to isolate `yprime`. The algebra here was so simple that we could do it faster in our head and writing down the answer. However, as the complexity of the equation increases, this calculator technique becomes useful.

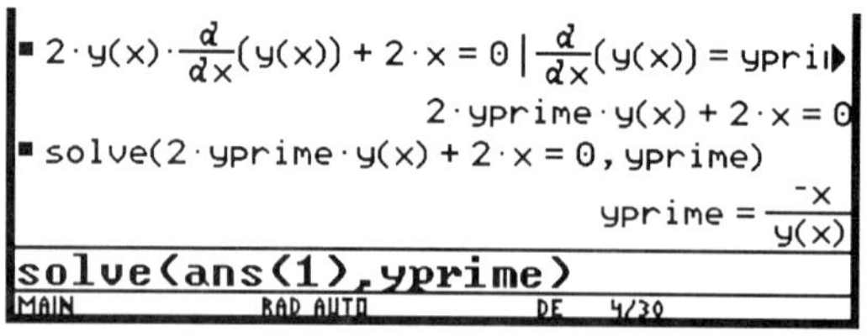

Figure B.3 `Solve` *for the derivative,* `yprime`.

➤ *Tip: Do not reverse the sides of the equality condition in step two. Should you use* `yprime=d(y(x),x)`, *no substitution will be made. Also be careful to use the derivative symbol* d, *not the letter* d.

PART III
INTEGRAL CALCULUS

13. LEFT- AND RIGHT-HAND SUMS
14. THE DEFINITE INTEGRAL
15. THE INDEFINITE INTEGRAL AND THE FUNDAMENTAL THEOREM OF CALCULUS
16. RIEMANN SUMS
17. IMPROPER INTEGRALS
18. APPLICATIONS OF THE INTEGRAL

DEMONSTRATIONS:
- A. PARTIAL FRACTIONS
- B. THE USE AND CONVERSION OF UNITS

13. LEFT- AND RIGHT-HAND SUMS

The fundamental activity of integral calculus is adding. In the discrete case, we sum a set of values. In the continuous case, we use the integral to sum over an interval. In this chapter we will restrict our attention to finite discrete sums. You probably already know that these sums are approximations to the value of a definite integral. We will make that connection in Chapter 16, Riemann sums.

Distance from the sum of the velocity data

Time (sec)	0	2	4	6	8	10
Velocity (ft/sec)	20	30	38	44	48	50

Figure 13.1 Velocity of a car every two seconds.

If you drive 50 miles per hour for 3 hours, then you'll have traveled 50 + 50 + 50 = 150 miles. We rarely travel a constant speed. If you drive 20 mph for two hours, 30 mph for the next two hours, and finally 40 mph for two more hours, then you will have traveled 20(2)+ 30(2) + 40(2) = 180 miles over the six hours.

Figure 13.1 shows velocity readings at 6 different times. We do not know if we traveled mostly at 20 ft/sec or 30 ft/sec for the first two seconds. If we assume the velocity is constantly increasing, then these two numbers give us lower and upper bounds for speed in the first two seconds. To get a lower bound on the distance traveled in the ten seconds, reason as follows. For the first two seconds, the velocity was at least 20 ft/sec, so the car traveled at least 40 feet in that interval. In the second two seconds, the car traveled at least 60 feet. A lower bound on the total distance traveled, then, is a sum of the first five velocities multiplied by two. For an upper bound, double the sum of the last five velocities.

The TI has list variables (named like any other variables) and list operations to do this kind of job. To distinguish list variables, press **2nd_VAR-LINK** and the variable names and types are shown in a scrolling window. You may want to review our earlier work with lists in Chapters 2 and 7.

Creating and summing lists from the home screen

We enter the first data list as shown in Figure 13.2. To create the second list, we edit the first list when it is left on the entry line. The **sum** command can

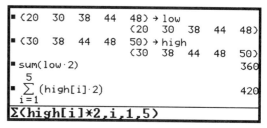

Figure 13.2 Finding lower and upper bounds on distance traveled.

be typed or pasted from the `2nd_MATH 3:List` submenu. A second way to sum a list is to use the Σ command (this symbol is the `2nd` version of the `4` key). This is the more complicated option, but it will be necessary later so we'll begin working with it now. The summation syntax is very close to the traditional mathematical notation.

$$\sum_{i=1}^{5} 2 \cdot high(i) = \Sigma(high[i]*2,i,1,5)$$

➤ *Tip: The notation* `high[3]` *means the third element in the list named* `high`*. You will get an error if* `high` *is not a list or if it does not have a third element. The number of elements in a list, its dimension, is given by* `dim(listname)`.

➤ *Tip: Some users like to start all their list variable names with L, but on the TI-92 this is translated to the lower case (* `l` *), which looks much like a one (* `1` *) in the calculator's font.*

Summing lists to create left- and right-hand sums: Σ

The previous example summed a simple list. A common calculus task is to form the left- and right-hand sums for a function over an interval that has been divided into n equal subintervals (where n is an arbitrary integer that tends to get larger and larger). This is only slightly more complicated than what we have done; it is still a sum of products of function values and the fixed length of an interval subdivision. Geometrically, it is the sum of areas of a collection of rectangles that are $f(x)$ high and Δx wide. That is, it approximates the area between the function's graph and the x-axis. In Chapter 16 we will use a program that draws the rectangles and calculates left- and right-hand sums; here we just want to understand how to build these sums.

The left- and right-hand sums

We will create the left-hand function and then make minor adjustments for the right-hand definition. First we will assume that the function whose values we sum is defined as `y1`. Sums are calculated over the interval $a \le x \le b$ divided into n of partitions, each of length $\Delta x = (b-a)/n$. The left-hand sum uses the function values at the left of each sub-interval, so the relevant x-values are

$a, a + (b-a)/n, a + 2(b-a)/n, \ldots, a + (n-1)(b-a)/n$ (so not including b).

We will call the sum function `lhs` and use the three letters `a,b,n` as independent variables. The most compact definition uses `STO→` (as opposed to `Define`).

Extra parentheses have been added to influence the history area format, shown Figure 13.3. (Remember that the function is stored as y1). The entry line for the definition is too long to show on the calculator screen:

```
Σ(y1(a+i*((b-a)/n)),i,0,n-1))*((b-a)/n),i,0,n-1)→lhs(a,b,n)
```

Once entered, the formula can be tested by using the values shown in Figure 13.3. The decimal answers can be obtained by pressing ◆_≈ (or by setting MODE to APPROXIMATE.) The right-hand sum differs by using the function values at the right of each sub-interval, so we start at $a + (b-a)/n$ and go to $a + n(b-a)/n = b$, i.e., the index runs from 1 to n instead of 0 to $n-1$. To enter this function, get the definition of lhs from the history area and make the following minor changes: the *start* changes from 0 to 1, the *stop* from n-1 to n, and 1 becomes r in the name. Notice that the right-hand sum for $1/x$ is less than the left-hand sum. The data in the velocity example indicated an increasing function, while $1/x$ is decreasing.

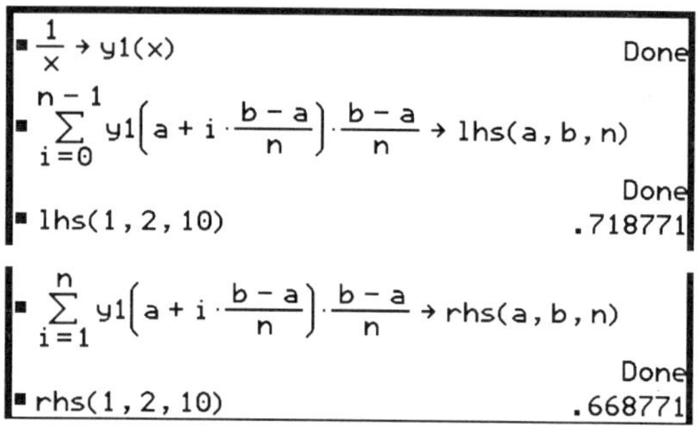

Figure 13.3 The left- and right-hand functions are defined and used to find sums of $f(x) = 1/x$ over $1 \le x \le 2$ with 10 subdivisions.

Negative values in the sum

In Figure 13.4 we consider the function $f(x) = \sin(x^2)$. The left-hand sum on the interval $0 \le x \le \sqrt{2\pi}$ is less than the left-hand sum on the shorter interval $0 \le x \le \sqrt{\pi}$. How can the area under the graph decrease by using a larger interval? The graph shows that the function is negative from $\sqrt{\pi}$ to $\sqrt{2\pi}$. The decrease in the sums now makes sense since we are adding negative values when $\sqrt{\pi} \le x \le \sqrt{2\pi}$; rectangles below the *x*-axis contribute "negative area." Thus we need to be careful when interpreting these sums as areas. To find the actual area between the graph and the *x*-axis, we would have to use the absolute values of any negative terms in the sum.

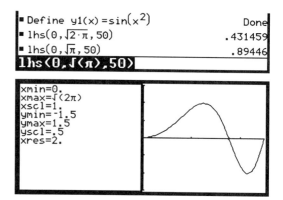

Figure 13.4 The total sum over an interval may be less than the sum over a subinterval if the function is negative.

Approximating area using the left- and right-hand sums

By increasing the number of partitions, the left- and right-hand sums may approach a limit which we interpret as the (signed) area under the function's graph. We write this as

$$\int_a^b f(x)\,dx = \lim_{n\to\infty} \sum_{i=1}^{n} f(x_i)\cdot \Delta x$$

```
■ Define y1(x)=sin(x)           Done
■ lhs(0,π,10)                 1.98352
■ lhs(0,π,50)                 1.99934
■ lhs(0,π,100)                1.99984
■ nInt(y1(x),x,0,π)                2.
```

Figure 13.5 The left-hand sum of the sine function on the interval $0 \leq x \leq \pi$ approaches the value 2.

(Recall that the dependence on a, b, and n is hidden in the definitions of x_i and Δx in the summation.)

As an example, let's see if there is a limit to the left-hand sums of $\sin(x)$ over the interval $0 \leq x \leq \pi$ as the number of partitions increases. The computations with $n = 10$, 50 and 100 in Figure 13.5 suggest that the sums approach 2. The TI-92 has a built-in function `nInt` that reports the numeric limit in these cases. We will later confirm that $\int_0^\pi \sin(x)\,dx = 2$ with the Fundamental Theorem of Calculus.

But we know from Chapter 7 that the TI-92 has a `Limit` function. So we apply it to this left-hand sum. Unfortunately, we see in Figure 13.6 that it does not evaluate the limit of the `lhs` function, although it does simplify the summing expression by setting $a = 0$ and $b = \pi$. There are infinite sums that the TI-92 knows and these can be used to verify that the definite integral is the limit of the left- (or right-) hand sums. Using `y1(x)=x^2`, we see that both sums have the same limit on the interval $0 \leq x \leq 1$. Again, we can confirm that this number is the value of the definite integral by using the Fundamental Theorem of Calculus.

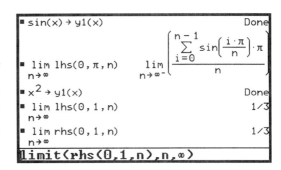

Figure 13.6 The `Limit` used to derive a value of the definite integral from the left- and right-hand sums.

➤ *Tip: The limits in Figure 13.6 were taken as $n \to +\infty$, but $\infty-$ is shown in the one symbolic outputs. This means that the limit is one-sided — all values are below, to the left of positive infinity, and the negative sign denotes this.*

➤ *Tip: Older TI calculators use the combination `sum(seq())` for Σ. In particular, `Σ(x^2,x,0,4)` corresponds to `sum(seq(x^2,x,0,4,1))`.*

14. THE DEFINITE INTEGRAL

We have just seen in the previous chapter that we can calculate left- and right-hand sums which approximate the signed area between a curve and the *x*-axis. The definite integral is defined as the limit of the left-hand (or right-hand) sum as the number of partitions goes to infinity. Thus each definite integral is a specific real number and the TI-92 will calculate this value or, in some cases, give an approximation that is generally reliable. The definite integral is evaluated as a number, but we will see that we can let its upper limit be a variable and thus create a new function. As with the derivative, values of the integral of a function can be found on both HOME and GRAPH screen.

The definite integral from a graph: ∫f(x)

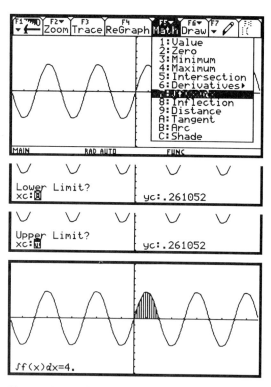

From a function's graph, we can find its definite integral *and* see the graphic representation. We start with $y = 2\sin(x)$ and graph it in Figure 14.1. As with finding the derivative from a graph, we use the F5 (Math) pull-down menu and its integral option 7: ∫f(x)dx. We are prompted to set the lower limit and then the upper limit. These limits are set in the same way that you have already set bounds using several other commands. Remember that the limits must be within xmin and xmax.

The number ∫f(x)dx can be interpreted as the area of the shaded region. It may surprise you that the result is an integer, but recall from the last chapter that the left- and right-hand sums of the sine function converge to 2 over this interval.

Figure 14.1 The definite integral of y1 *=*2sin(x) *with* $0 \leq x \leq \pi$ *and the* ZoomTrig *setting.*

➤ *Tip: If there is more than one function graphed, then you must choose the desired function after selecting the* ∫f(x)dx *command, before specifying a lower limit.*

► *Tip: There are four styles of shading that are used consecutively, so your shading styles may differ from the ones shown in the examples.*

The definite integral as a number on the HOME screen

To find the same result without using a graph, you can work from the HOME screen. The TI-92 command for the definite integral is the yellow integral sign, the 2nd version of the 7 key. It is also the second option of the F3 (Calc) pull-down menu. The entry line command in Figure 14.2 shows the syntax used to produce the answer and traditional mathematical symbolism in the history area: after the function and variable, give the lower and upper limits of integration.

Figure 14.2 HOME screen integration.

Facts about the definite integral

Four definite integral facts will be illustrated below using simple functions and their graphs. You are encouraged to change the function and window to up the excitement. In each of the following examples, we show the result graphically and then, in the final frame, the numerical rendition of the same result on the HOME screen. It is important that you feel comfortable using both methods of finding the definite integral.

► *Tip: After a graph with shading has been displayed, you will usually want to clear the screen before the next graphing. This can be done with F4 (ReGraph).*

Reversed limit integrals are the negative of one another

Unlike most parameters, which prompt error messages whenever xmin > xmax or Lower Bound > Upper Bound, the Upper Limit and Lower Limit can be in either order. Reversing the order changes the sign of the answer, as demonstrated in Figure 14.3.

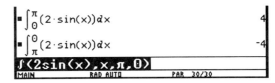

Figure 14.3 Limit reversal changes the sign of the definite integral.

► *Tip: The graph method is not shown here because shading does not indicate the reversal of the limits (although the ∫f(x)dx values do switch sign).*

The intermediate stop-over privilege

The definite integral can be calculated as a whole from the lower to upper limit, or it can be calculated in contiguous pieces. This can be thought of as a plane fare where the charge is the same whether you fly non-stop or have intermediate landings. In Figure 14.4 we get the same answer as before by considering our function y1=2sin(x) over $0 \leq x \leq \pi/3$ and then $\pi/3 \leq x \leq \pi$.

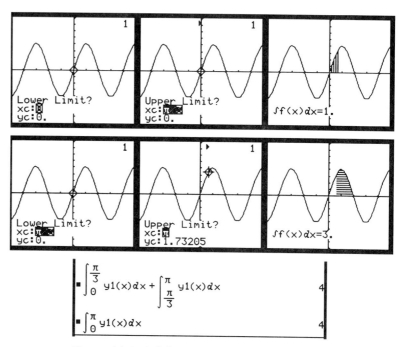

Figure 14.4 *A definite integral found in two pieces. (The graphs are halves of split screens.)*

The definite integral of a sum is the sum of the integrals

In Figure 14.5, on the opposing page, we have set y1=2sin x and y2=x. The graph of the function y3=y1(x)+y2(x) is shown as a Thick curve. We see, both graphically and symbolically, that the definite integral of the sum function y3 is the sum of the definite integrals of the two summand functions.

Constant multiples can be factored out of a definite integral

We saw an example of this fact earlier with $\int_0^\pi 2\sin(x)dx = 2\int_0^\pi \sin(x)dx$. Another example is shown in Figure 14.6. We first check numerically to see that the two integrals are the same, then we find $\int_0^\pi 3\sin(x)dx = 6$ from the graph (which also shows sin(*x*) for comparison).

14. THE DEFINITE INTEGRAL   **103**

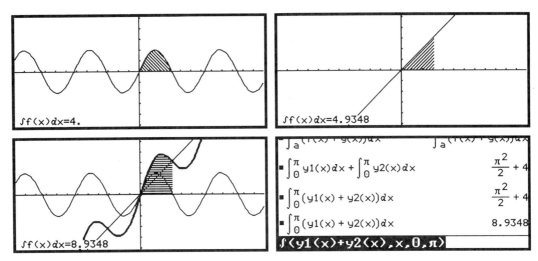

Figure 14.5 The definite integral of a sum is the sum of the integrals. Graph settings are ZoomDec. *When computing the definite integral of* y3, *you will need to use the up arrow until a 3 appears in the upper right corner to have the TI-92 refer to that function. Remember that* ♦_ ≈ *forces an approximate answer.*

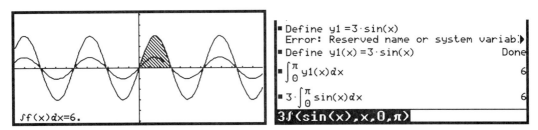

Figure 14.6 A constant multiple of a function can be factored out of a definite integral. In the HOME *screen, we committed and corrected a common error.*

The definite integral as a function: y=∫(...,x)

In this chapter's integral examples, we have used x as the function variable. We could have used some other dummy variable in the definite integral command and the numeric answer would be the same. For example, ∫(sin(t),t,0,π) would also give 2.

We usually think of the definite integral as a number resulting from the command ∫(f(t),t,a,b). Now if we change b, we get a new number... This is a function! As an example, use $\cos(t)$ as the function, choose 0 for the lower limit, and in the Y= editor define

$$y1 = ∫(\cos(t),t,0,x)$$

You may be wondering how we can do this: doesn't x have to be the variable for functions in the Y= editor? The t in the integral is a dummy variable; the answer to the definite integral depends on *x*, so y1 ultimately is a function of *x*. The graphing shown in Figure 14.7 is unbelievably slow because of the intensive numerical work to calculate each definite integral, so much so that the status line displays the warning questionable accuracy. The graph of y1 looks like a sine function. To further investigate, we define y2=sin(x) and compare table entries. (You could also graph the sine function and check that the two functions have the same graph.)

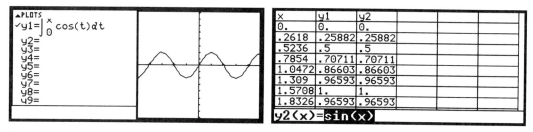

Figure 14.7 *The integral function of the cosine appears to be the sine function. (Settings are* ZoomTrig *for the graph,* tblStart = 0 *and* △tbl = π/12 *for the table.)*

▶ *Tip: Graphing functions defined with* ∫(...,x) *is quite slow; setting* xres *higher will increase graphing speed.*

The TI-92's computer algebra system will verify this for us. On the home screen in Figure 14.8, the entry ∫(cos(t),t,0,x) gives sin(x). We also see that interesting things happen when we change the lower bound from zero. The three functions defined this way differ by constants, so graphically they are vertical shifts of one another. Thinking of the derivative as the rate of change of *y*-values, these three should have the same derivative: each of these three functions is called an antiderivative of cos(*x*). The common notation used is

$$\int \cos(x)\,dx = \sin(x) + C$$

where *C* is an arbitrary constant. This will be the focus of the next chapter.

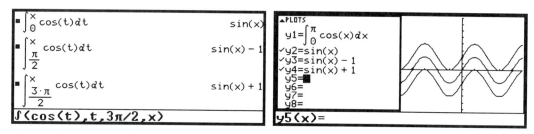

Figure 14.8 *Integral functions of the cosine starting with different initial values are of the form* sin(x) + C. *Remember that graphing functions with known formulas is much faster than using an integral definition.*

15. THE INDEFINITE INTEGRAL AND THE FUNDAMENTAL THEOREM

The Fundamental Theorem of Calculus will be discussed in two contexts: as the total change of the antiderivative, and as a connection between integration and differentiation. In the last chapter we saw, for a particular function, that the definite integral with variable upper limit changes by only a constant when the lower limit is changed.

The antiderivative

The antiderivative of a function is found symbolically in either of two forms. We see in Figure 15.1 that ∫(cos(x),x) gives the antiderivative with constant zero, while ∫(cos(x),x,c) adds the constant. It should be noted that some functions have no analytically defined antiderivative. When the antiderivative is not known, the output will be the original input with as much processing as could be done.

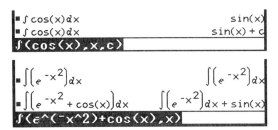

Figure 15.1 Two forms of finding the antiderivative. When the antiderivative cannot be found, the integral notation will be shown as the output.

The Fundamental Theorem

The Fundamental Theorem of Calculus states:

if f is a continuous function and $f(t) = \dfrac{dF(t)}{dt}$, then $\int_a^b f(t)dt = F(b) - F(a)$

One reason for using this theorem is that it calculates the definite integral in a simple way. Unfortunately, this use is limited to occasions when we know the antiderivative. For example, we previously used the Riemann sum to guess that the definite integral $\int_0^\pi \sin(x)dx = 2$; we confirm this using an antiderivative and the Fundamental Theorem. We could even do this calculation in our head:

an antiderivative of $f(x) = \sin(x)$ is $F(x) = -\cos(x)$,

so $\int_0^\pi \sin(x)\,dx = F(\pi) - F(0) = -\cos(\pi) - (-\cos(0)) = -(-1) - (-1) = 2$

The calculator uses the Fundamental Theorem in its symbolic answers. The first entry shown in Figure 15.2 verifies that the TI-92 knows an antiderivative of x^2. Then it uses the Fundamental Theorem with $F(x) = x^3/3$ to evaluate the integral symbolically as $F(b) - F(a)$.

Figure 15.2 The TI-92 uses the Fundamental Theorem to symbolically evaluate a definite integral.

The definite integral as the total change of an antiderivative

A second reason to use the Fundamental Theorem is that it gives us an exact answer, which may be required or just plain useful. For example, when a growth factor compounds continuously, the decimal accuracy is limited to that of the calculator. This is fine when we are dealing in thousands or millions, but sometimes we have amounts that are astronomical and we want an answer that will be exact to whatever number of decimal places are required. Think of the value of π: it is roughly 3, or, if more accuracy is needed, we can use 22/7, or better yet, 3.1459. In its exact form, the symbol π represents full accuracy, not a decimal or fraction approximation.

Let's look at an example where an exact answer will be found. Consider a savings account into which you put a dollar every hour. What will it be worth in 20 years if it is compounded continuously at a 10% annual rate? This is a thinly disguised definite integral. First, whatever you deposit needs to be expressed in an annual amount so that all our rates are annual. Call this amount P, which we will take as 365*24 (ignoring leap years). Deposits are so frequent that we will consider the rate to be continuous. The future balance in ten years is then given by the definite integral

$$\int_0^{20} Pe^{0.1(20-x)} dx$$

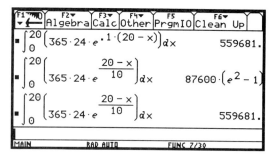

Figure 15.3 A definite integral calculated using the closed form given by the Fundamental Theorem. When decimals are involved the approximate answer is shown.

This gives a value of over half a million dollars. This is probably good enough in this case, but the calculator answer does have limited accuracy. Rewriting the equation with no decimals, we see in Figure 15.3 that the Fundamental Theorem is used to write an exact answer (correct to an infinite number of digits). The status line shows that we are in the **AUTO** mode of

15. THE INDEFINITE INTEGRAL AND THE FUNDAMENTAL THEOREM

evaluation, which works in the **EXACT** mode (also called *closed form*) whenever possible. When written in terms of *e*, the expression has full accuracy. Knowing the closed form solution allows us to accurately find the future value of saving a thousand dollars per minute, although we would need a calculator with more internal digits of accuracy to see the difference. The two answers are the same when compared by finding the approximate value of the second integral by using ♦ _ ≈.

➤ *Tip: The* **MODE** *setting can be changed from* **AUTO** *to* **EXACT** *to force all the integrations to be exact. The setting is shown on the status line. It is better to not to change mode settings unless absolutely necessary.*

Viewing the Fundamental Theorem graphically

In the above examples, the lower and upper limits were constants. Now, as we did in the last chapter, consider the upper limit as a variable. Specifically, replace *b* with *x* on each side of the Fundamental Theorem equation, giving a function in terms of *x*:

$$\int_a^x f(t)dt = F(x) - F(a)$$

We now consider an example by setting up the following functions:

$f(x) = $ **y1** $ = $ **x^2/10**, whose simplest antiderivative is $F(x) = $ **y3** $ = $ **x^3/30**

Since *a* is arbitrary, let *a* = -3.5. (So the integral is zero at *x* = -3.5: there is no width for the area there.) Finally, we define the two functions we want to compare and see if they are equal:

y2 = ∫(y1(t),t,-3.5,x) and **y4 = y3(x)-y3(-3.5)**

Using a **ZoomDec** graphing window (with **xres=2**) and graphing styles **y1** (**Thick**), **y2** (**Dots**), and **y4** (**Path**), you see in Figure 15.4 that **y2** and **y4** have the same graph.

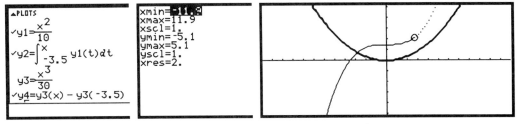

Figure 15.4 An example of the Fundamental Theorem with f(t) = t²/10.

Checking on f(x) = e^x, the function that is its own antiderivative

As another example of this kind of comparison, we let $f(x)$ be the famous exponential function, whose antiderivative is itself. We see in Figure 15.5 that the general exponential function is its own antiderivative (up to a constant coefficient).

Figure 15.5 *The antiderivative of the exponential function.*

This is why the number e is so important: the exponential function with base e is exactly its own derivative and antiderivative.

Comparing d(∫(...) ...) and ∫(d(...) ...)

What happens when you find the derivative of the integral function? It should not be too surprising that you get the original function back — or do we? In Figure 15.6 we try this out with $y = \sin(x)$ and we confirm that it is true for this function. Disregarding possible differences in constant terms, differentiating and integrating are inverse operations.

But now consider $g(x) = \sin(x)/x$. Even though the TI-92 does not give a symbolic answer to the integral of $g(x)$, it gives the original function for the derivative of the integral. However, this does not work for the integral of the derivative. This is another example beyond the calculator's symbolic integration capacity: it knows no antiderivative for either term of $g'(x) = \cos(x)/x - \sin(x)/x^2$. We can verify

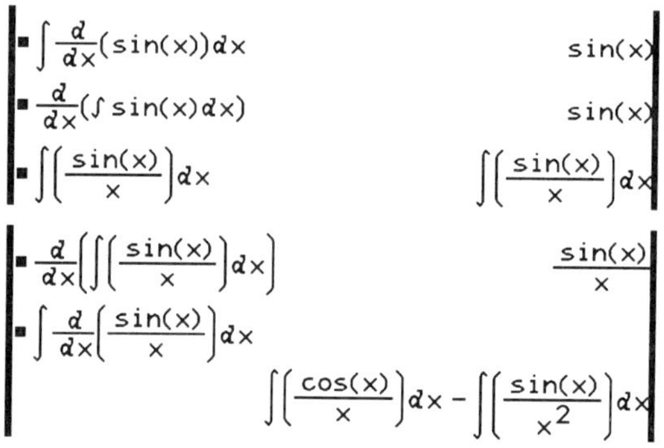

Figure 15.6 *Investing the derivative of the integral and vice versa.*

that the integral of $g'(x)$ really is $g(x)$ by graphing. To graph the integral, we will use t as the internal variable and x as the upper bound. The choice of the lower bound effects the constant of integration, so the integral will be verified if the two graphs are vertical shifts.

Graphing comparison

We are now faced with practical considerations of graphing. It would be cleanest to compare `y1=∫(d(sin(t)/t,t),t,1,x)` and `y2=sin(x)/x`, but even with `xres=5` the graphing of the first function is insufferably slow. Entering the formula for $g'(t)$ helps. The best option is using the F3 (Calc) option B:nInt instead of ∫; this speeds up the graphing by about a factor of three. Even so, the graph in Figure 15.7 took about three minutes. This verifies that the integral of the derivative is the original function, up to a constant.

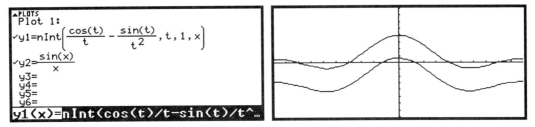

Figure 15.7 Graph showing that the (numeric) integral of the derivative differs from the original function by a vertical shift of -sin(1) = -0.84147... (the WINDOW *setting is* ZoomTrig *after using* ZoomIn *with a factor of 2).*

nInt versus ∫

In Chapter 9, when confronted with the rare function that the TI-92 could not differentiate symbolically, we used nDeriv to give an approximate function answer. The analog nInt is not as useful.

There are many functions that the TI-92 cannot integrate symbolically, for example $f(x) = e^{\wedge}(x^2)$. This is not a deficiency of the calculator, just a fact of calculus — many seemingly benign indefinite integrals have no closed form solutions. When the calculator is confronted with a definite integral, it checks to see if it knows the exact formula and, failing that, switches to a numeric method. Thus, unless you want to save the short time it takes in checking for an exact formula, you can use ∫ almost all the time. If the MODE is APPROXIMATE then most exact techniques are ignored. In cases where you know all the work will be done with numeric approximations, as with the graphing above, use nInt. The numeric methods for both of these integral commands are similar to the ones discussed in the next chapter.

Figure 15.8 A function without an exact form integral will automatically use nInt *to calculate the definite integral.*

16. RIEMANN SUMS

In Chapter 13 we introduced the left- and right-hand sums to approximate the definite integral. In this chapter we use a program to simplify explorations of these and other types of sums. We will add the capability to graphically view the subdivision areas that comprise the approximation.

A few words about programs

This marks our first use of a stored program, so perhaps an introduction is in order. A program is a set of commands that are performed in a prescribed order, like a recipe. The order is normally the sequential list of commands, but there are techniques to alter that order. Special program command menus (I/O and CTL) list the input/output and control commands.

A program is written by pressing APPS 7:Program Editor 3:New... and arrowing down to enter a program name in the Variable: box. The first choice (Type:) in the dialog box allows you to choose between a program and function, the second choice (Folder:) lets you assign a folder to work from. In this book, all work is done in the default folder named MAIN. If you want edit the last program you have used, then press APPS 7:Program Editor 1:Current and the current program will be put on the screen for editing. (Once you start editing a program, the key sequence APPS 7 1 allows for quick entry to the listing.) A colon begins each new program line. Use 2nd_QUIT to exit the PROGRAM EDITOR.

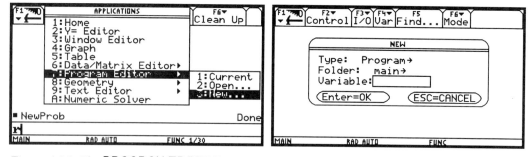

Figure 16.1 The PROGRAM EDITOR is an option in the APPS menu. A dialog box allows you to name a new program.

▶ *Tip: Use names of at least two characters for programs so that they are not destroyed when you periodically clear the single letter variables. You cannot store a number in a variable with the same name as a program.*

A program is activated — the more common terms are *run* or *executed* — by typing its name on the entry line and pressing ENTER. Step-by-step instructions can be found in the programming Chapter 17 of the TI-92 *Guidebook*.

Entering a program from the printed page is quite tedious and you can expect to make a few errors that will only be discovered when executing the program. However, once a program is correctly entered, it can be transferred to other TI-92 / TI-92 Plus calculators using the built-in Link features. In the classroom, it is typical that a program is verified by the instructor and distributed to the class using 2nd_VAR-LINK F3 (Link) 1:Send. You can also download programs from the Web; this is outlined in the Appendix.

Setup programs

Before we use a more complex program, you can practice creating a very simple one that automates a multi-step task. If you use the split screen feature of the TI-92, you have found that there is considerable navigation involved because Split Screen is on the second page of the MODE menu. In addition, it is convenient to have the two screen applications specified. In Figure 16.2 the sequence defines a program ss() that will split the screen and assign applications. Selecting the

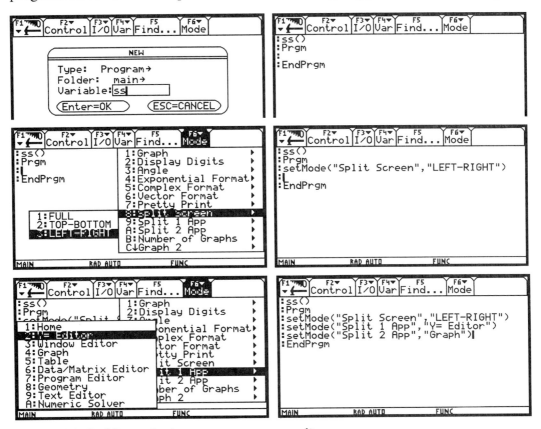

Figure 16.2 Building a simple program to set up a split screen.

Split Screen option LEFT-RIGHT does not change the screen setting at this time, but instead pastes a line of code that will split the screen when the program is used. (This process is similar to creating "macros" in some software packages.)

➤ *Tip: After pressing F6 (MODE), pressing the right arrow will show the submenu for an item even though the submenu will appear on the left. The left arrow would move to the F5 pull-down menu.*

Program entry error

The most common error when you first use programs is to forget that they are defined as functions and require parentheses (whether or not there are any input variables). A second common error is to not start the program name on an empty entry line.

Using the rsum() program to find Riemann sums

A program to automate the Riemann sum process is given at the end of this chapter. It is designed to require a minimum of input within the actual program.

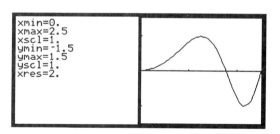

Figure 16.3 Set y1=sin(x^2) and the window shown before using the rsum() program.

Before using it, you must define your function in y1 and set the window to have xmin be the left endpoint and xmax be the right endpoint of the desired interval. In Figure 16.3 shows these preliminary steps.

In Figure 16.4 we first see the program name rsum() pasted from the submenu and started from the home screen. A few moments after activation, the program prompts you for the number of subdivisions (partitions) of the interval. Next, five different kinds of Riemann sums are calculated and displayed. The value labeled nInt is short for nInt(y1,x,xmin,xmax) and can be used to judge the other numerical approximations. The last value, labeled Simpson, is the value given by Simpson's method. This is a weighted average of the two previous results, specifically the sum of the trapezoid value and twice the midpoint value all divided by three. You will find that this value is consistently close to the 'true' value for small partitions. (The nInt value comes from a more complex weighted average.)

16. RIEMANN SUMS   113

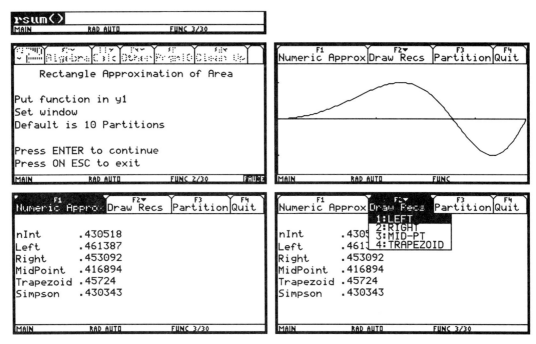

Figure 16.4 The program rsum() *gives six numeric approximations of the definite integral of* y1 *from* xmin *to* xmax.

In Figure 16.5, the graphic representations of the Riemann sums are shown in the order of the four menu choices on the menu line. Pressing **F3** (**Partition**) will allow you to reset the number of partitions. Pressing **F4** (**Quit**) will exit from the program.

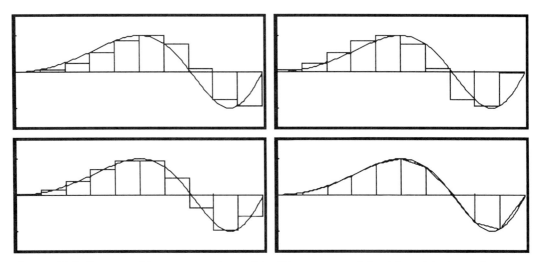

Figure 16.5 Four graphic representations of the Riemann sum: left, right, midpoint and trapezoid.

▶ *Tip: When you exit a program, you may be left in a variety of screens. Use* ♦_HOME *to return to the* HOME *screen.*

The TI-92 program rsum()

The program listed below takes a considerable amount of time to enter, but once entered, can be passed to other TI calculators of the same type. (It is in the public domain.) It can also be stored on a computer using the TI-Graph Link™.

If you are entering this by hand, you can use a copy-and-paste technique (♦_C & ♦_V) on repetitive code. The Lbl a1 through Lbl a5 sections are similar, so paste in four copies of the section and edit them in the few places where they differ. Indentation and spaces are permitted; they can make program code more readable. Special symbols such as ≥ are found in the 2nd_CHAR menu. The numeric approximation sums use the sum(seq(...)) format to stay closer to the TI-Eighty calculator programs published in the companions to this book. Once you have this program working, if you are feeling adventurous, you might modify it to accept more inputs and use more TI-92 specific features.

```
rsum()
Prgm
ClrIO
Disp "    Rectangle Approximation of Area"
Disp " "
Disp "Put function in y1"
Disp "Set window "
Disp "Default is 10 Partitions "
Disp " "
Disp "Press ENTER to continue"
Disp "Press ON ESC to exit"
Pause
FnOff
10→n
ClrDraw
DrawFunc y1(x)
StoPic pic1
nInt(y1(x),x,xmin,xmax)→u

Lbl a0
(xmax-xmin)/n→d
ToolBar
 Title "Numeric Approx",a1
 Title "Draw Recs "
  Item "LEFT",a2
  Item "RIGHT",a3
  Item "MID-PT",a4
  Item "TRAPEZOID",a5
 Title "Partition",a6
 Title "Quit",a7
EndTBar
```

```
Lbl a1
ClrIO
Output 10,1,"nInt"
Output 10,60,u
Disp "Left"
sum(seq(y1(xmin+i*d),i,0,n-1,1))*d→l
Output 21,60,l
Disp "Right"
sum(seq(y1(xmin+i*d),i,1,n,1))*d→r
Output 32,60,r
Disp "MidPoint"
sum(seq(y1(xmin+i*d),i,0.5,n,1))*d→m
Output 43,60,m
Disp "Trapezoid"
(l+r)/2→t
Output 54,60,t
Disp "Simpson"
(2*m+t)/3→s
Output 65,60,s
Goto a0

Lbl a2
RplcPic pic1
For i,0,n-1
xmin+i*d→x
y1(x)→y
Line x,0,x,y
Line x,y,x+d,y
Line x+d,y,x+d,0
EndFor
Goto a0

Lbl a3
RplcPic pic1
For i,0,n-1
xmin+i*d→x
y1(x+d)→y
Line x,0,x,y
Line x,y,x+d,y
Line x+d,y,x+d,0
EndFor
Goto a0

Lbl a4
RplcPic pic1
For i,0,n-1
xmin+i*d→x
y1(x+d/2)→y
Line x,0,x,y
Line x,y,x+d,y
Line x+d,y,x+d,0
EndFor
Goto a0

Lbl a5
RplcPic pic1
For i,0,n-1
xmin+i*d→x
y1(x)→y
```

```
Line x,0,x,y
Line x,y,x+d,y1(x+d)
EndFor
Line xmax,y1(xmax),xmax,0
Goto a0

Lbl a6
ClrIO
Input "Partitions:",n
If n≥0
Goto a0

Lbl a7
ClrDraw
ClrIO

EndPrgm
```

17. IMPROPER INTEGRALS

In this chapter we will look at two different but similar problems encountered when trying to use the integral in a wider setting. These special cases that involve infinity are called improper integrals. First, we will see that sometimes the limits of integration can be infinite. Second, we will see that integration is sometimes possible even when the integrand function itself has infinite values.

An infinite limit of integration

On a first take, you might think that any positive function that goes on forever must have an infinite definite integral. Let's try a thought experiment. Suppose you decided to go on a diet and every day you cut your chocolate chip cookie consumption in half. How many cookies would you need for your lifetime? (or for eternity?) Visualize the cookie: the first day you would eat half, the next day a half of a half (a quarter), and so on. Because each day you would only eat half of the remaining cookie, you would never finish it: one cookie would last a lifetime! Figure 17.1 shows the sum is 1. (The ∞ symbol is the **2nd** version of the **J** key.)

Figure 17.1 *The sum of cookie halving is one.*

Three power functions

Let's compare the three functions

$$y_1 = \frac{1}{x}, \ y_2 = \frac{1}{x^3} \text{ and } y_3 = \frac{1}{x^{1/3}}$$

as x goes from 1 to infinity and see if we get a graphical hint about which might have a finite area under it. For this to happen, the function must approach zero. We see in Figure 17.2 that all three functions do approach zero as x gets large, and y_2 does so the fastest. Just like the sum in Figure 17.1, we can find the improper integral by using infinity as the upper limit.

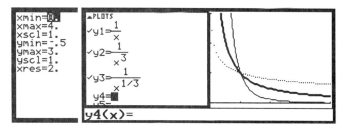

Figure 17.2 *Comparing three functions as x goes to infinity. Styles are* **Thick** *for* **y1**, **Line** *for* **y2**, *and* **Dot** *for* **y3**.

In Figure 17.3 we see that the integral for y2 converges, but not those for y1 and y3. However, these divergent integrals are assigned finite values in the APPROX mode. This is because the calculator's numeric method erroneously decided at some point that it could ignore the decreasing terms in its area summation. So we should pay attention to the evaluation mode and question results that are done in the APPROX mode. If your functions involve decimals, then you will need to change the MODE setting from AUTO to EXACT to be sure of answers to improper integrals.

▶ *Tip: When evaluating improper integrals, expect long calculation times. There is no way good way to speed this up. Should the wait be too long to bear, press ON to halt the computation.*

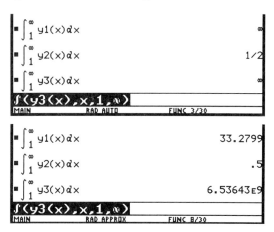

Figure 17.3 *Evaluating improper integrals with* EXACT *and* APPROX *settings, the latter giving incorrect answers.*

The whole story on $\int_1^\infty \frac{1}{x^p} dx$

An important family of functions is the negative power functions. We saw above that the integral diverged when $p = 1$ and converged when $p = 1/3$. We can further investigate this symbolically with the TI-92. In Figure 17.4, we see the results of the general improper integral: undef means that the answer cannot be determined. When we qualify the values of p, we get the answer we seek. You can investigate other examples such as $1/x^2$ and $1/x^{1/4}$ to test the following analytic results:

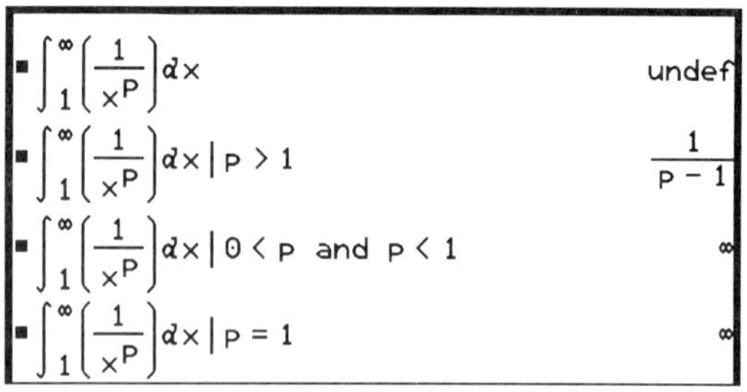

Figure 17.4 *The negative power function diverges when $p \geq 1$.*

If $p < 1$ then $\int_1^\infty \frac{1}{x^p} dx$ converges

If $p \geq 1$ then $\int_1^\infty \frac{1}{x} dx$ diverges

The convergence of $\int_0^\infty \frac{1}{e^{ax}} dx$, $a > 0$

It is simple to show analytically that $\int_0^\infty \frac{1}{e^{ax}} dx = \frac{1}{a}$ for $a > 0$. Let's try this out using a as an undefined variable. The first line of Figure 17.5 has an entry error: no multiplication sign was inserted between a and x. The calculator understood ax as an independent variable and gave an answer based on $\int_0^\infty dx$. With that corrected, we find as in the previous case that a needs to be restricted. The last entry gives the result we are expecting.

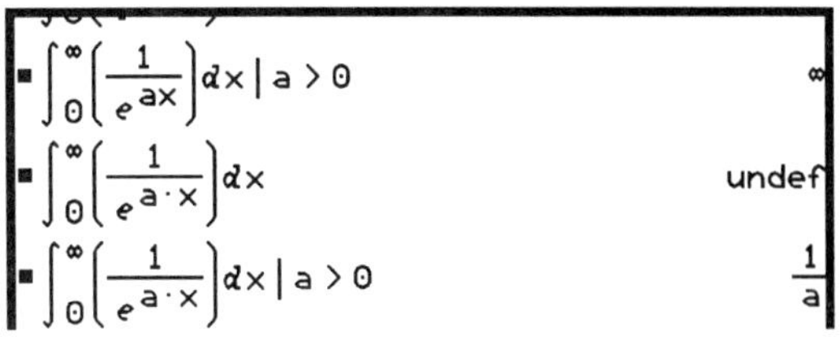

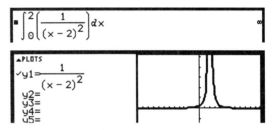

Figure 17.5 Symbolic evaluation of an improper integral beginning with a mistake.

The integrand goes infinite

The second major problem that makes an integral improper is an infinite integrand (one version of being undefined). This is a much more dangerous situation because there is no ∞ symbol in the integral to alert you. It is a good habit to graph the function before finding the definite integral. The graph should alert you to potential problems, like the integrand being undefined (and tending to positive or negative infinity). We show this in Figure 17.6, where we blindly find the definite integral with the value of the integrand being infinite at $x = 2$. By drawing a graph, we can see why the integral diverged. However, Figure 17.7 shows an example of an improper integral that converges even though the interval includes a point where the integrand goes to infinity.

Figure 17.6 An improper integral with finite limits and an infinite integrand that diverges.

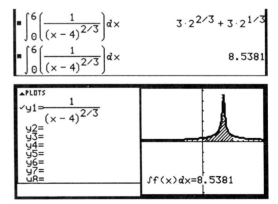

Figure 17.7 An improper integral with finite limits and an infinite integrand that converges.

Results that make us wonder

We have seen that the integral function may give misleading (incorrect) results when using APPROX (either with the MODE setting or ◆_≈). There is another danger: when a function is not defined. In Figure 17.8, we attempt to integrate over an interval not in the function's domain. In one case, the calculator gives an answer which we may not recognize as non-real (the TI-92 without Plus gives undef instead). In the second (very similar) case, it is more helpful by giving us an error message that the answer is non-real.

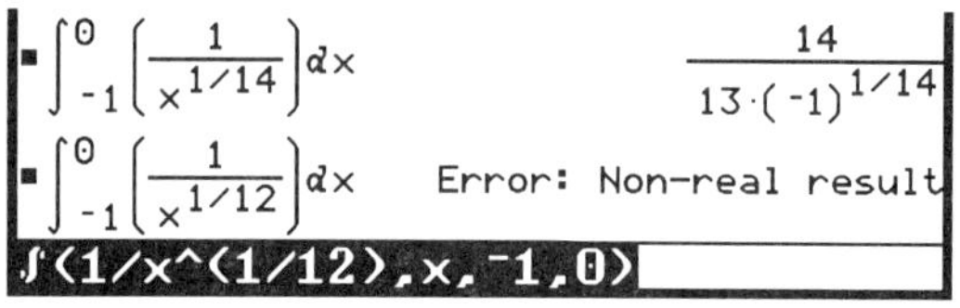

Figure 17.8 Strange answers for integrals over intervals outside the functions' domain.

A calculus teacher can easily make up examples where the calculator will mislead you. Would someone really do that? Yes.

The comparison test

The negative power function and the negative exponential function are of special importance because they are often used as a benchmark for comparison with more complicated functions. For example, suppose we wanted to know if the integral

$$\int_4^\infty \frac{1}{\ln x - 1} dx$$

converges or diverges. Entering this on the home screen (as shown in Figure 17.9) results in a long wait, a very large number as an answer, and the status line warning questionable accuracy. Graphically comparing this function y2 to y1=1/x, we see that y2 is above y1 in the window ($4 \leq x \leq 1000$). Knowing the long-term behavior of the natural logarithm we can conclude that y2 > y1 for all $x \geq 4$. Since the integral for $1/x$ diverges, we deduce that the larger integral in question must also diverge. The status warning questionable accuracy should not be ignored in this case.

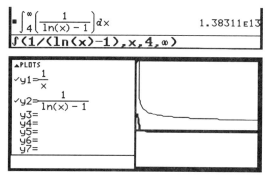

Figure 17.9 A shaky symbolic result and a graph showing that y2 (Line) exceeds y1 (Thick) on $4 \leq x \leq 1000$.

➤ *Tip: Watch the status line when doing integration. The warning* questionable accuracy *should be put you on guard.*

18. APPLICATIONS OF THE INTEGRAL

In this chapter, we look at four diverse applications of the integral. These standard examples give only a flavor of the extensive applications of this core mathematical concept.

Geometry: arc length

The following calculus formula finds arc length along a function's graph.

$$\text{Arc length} = L = \int_a^b \sqrt{1 + (f'(x))^2}\, dx$$

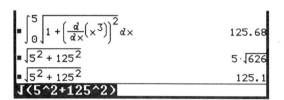

Let's calculate the arc length of the curve $y = x^3$ from $x = 0$ to $x = 5$ and compare it to direct distance from the origin to the point (5, 125). In Figure 18.1, we see this result calculated on the home screen.

Figure 18.1 Comparing arc length (given by the integral) and the distance between the endpoints.

The TI-92 will calculate both arc length and distance between points on a displayed graph. From the **F5 (Math)** menu, select **B:Arc** or **9:Distance**. The usual bound prompts appear, this time as **1st Point?** and **2nd Point?**. Both results are shown in Figure 18.2, where you see the distance line will be drawn on the graph for the **Distance** computation. The graph's scale makes the difference between the two lengths look greater than the actual numerical difference.

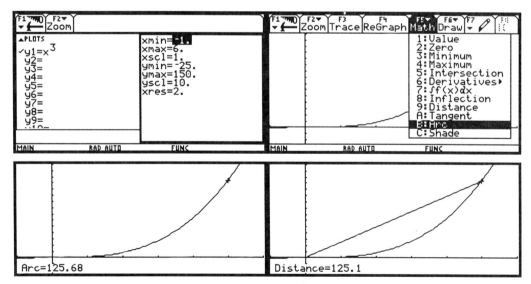

Figure 18.2 Finding the arc length of the curve $y = x^3$ from $x = 0$ to $x = 5$ and comparing it to the direct distance between (0, 0) and (5, 125).

Physics: force and pressure

Pressure increases with depth so that there is more pressure at the bottom of a container than at the top. One cubic foot of water weighs 62.4 pounds and exerts a force of 62.4 pounds on the base of its container. With half a cubic foot, the force is 62.4/2 = 31.2 pounds on the base. The pressure on the base is directly proportional to the depth of the water. We also know that force is the product of pressure and area.

The difficult part of pressure, force, or volume problems is not the actual integration that is required, but setting up the integral to accurately reflect the geometry of the situation. A time-honored system is to write the pressure as a sum of forces acting on strips or slices.

The pressure on a trough

Consider the trough shown in Figure 18.3. There are four sides that the force of water acts on. Let's tackle the easiest side first: the 3' by 14' horizontal back. We subdivide the height into pieces of length Δh, so this back side is made up of horizontal strips, each having an area of $14 \cdot \Delta h$ square feet. The entirety of each strip is at the same depth, so that all along a given strip there is equal pressure, namely

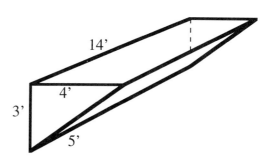

Figure 18.3 *A trough with the shape of a triangular prism.*

$$\text{force on a horizontal strip} = (62.4h)(14 \cdot \Delta h)$$

(We could multiply the constants, but the calculator can do that and we choose to leave our setup in this more readable form.) The total force is sum of all the horizontal strips. The exact total is given by the definite integral

$$\int_0^3 (62.4h)(14)dh$$

Next we consider the inclined side. For a Δh (vertical height change) the actual width w on the inclined side is greater than Δh. Using similar triangles, these values are related by $\Delta h/w = 3/5$. The area of the strip is $14(5/3)\Delta h$, so

$$\text{force on inclined strip} = (62.4h)(14)(5/3)\Delta h$$

The total force is given by the integral

18. APPLICATIONS OF THE INTEGRAL

$$\int_0^3 (62.4h)(14)(\tfrac{5}{3})\,dh$$

Finally we compute the force on the two triangular ends. Again using similar triangles, we find the area of a strip is $(4/3)(3 - h)\Delta h$ and

$$\text{force on end strip} = (62.4h)(4/3)(3 - h)\Delta h$$

so that the third integral is

$$\int_0^3 (62.4h)(\tfrac{4}{3})(3-h)\,dh$$

Now the easy part is entering these definite integrals for evaluation and computing the final sum; this is done in Figure 18.4.

➤ *Tip: It is much safer to leave calculator results in an expanded form so that the derivation remains evident.*

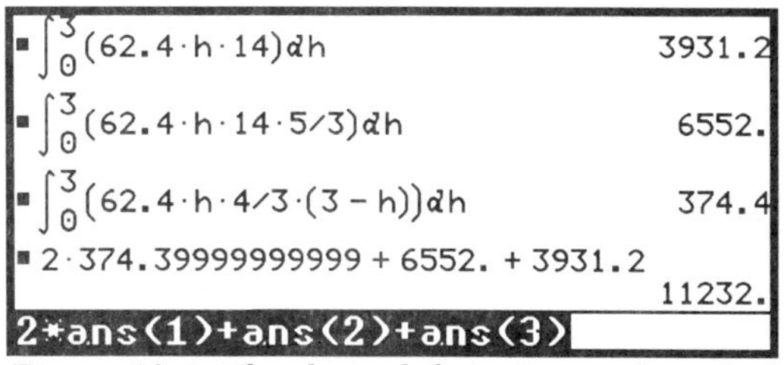

Figure 18.4 *The three definite integrals to find the force on a trough. (The TI-92 without Plus does not store the .4 as .3999...)*

Economics: present and future value

Suppose you win a two million dollar lottery. Before you spend the money, you are told that the money will be distributed to you over the next twenty years. That is a mere $100,000 per year (before taxes). If you wanted to get an immediate lump sum, you could sell your rights to all the future payments. What is this really worth? In economics this value is called the present value, V. It can be calculated using a fixed investment rate and an integral. If P is the annual payment and r is the annual investment rate for T years, then the present value is

$$V = \int_0^T Pe^{-rt}\,dt$$

Suppose that the agreed investment rate was 10% for the twenty years. Then we enter the integral as shown in Figure 18.5. The present value is $864,665, less than half the advertised amount!

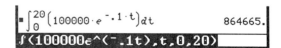

Figure 18.5 *The present value of a two million dollar lottery. Payments are $100,000 per year.*

Using `Numeric Solver` for "what if" analysis with `nInt`

If you have a TI-92 Plus, choose from the APPS menu `A:Numeric Solver` to set up an interactive session that allows us to both evaluate the present value and see the effect of different investment rates. Since the calculator reads all variables as lower-case letters, we use y in place of T for years. We also use `nInt` rather than `∫`, since it is better suited for looking solely at numeric values.

After entering the equation, we check our setup by recalculating the present value of twenty annual payments of $100,000 with an investment rate of 10%. To do this, enter `p=100000, r=.1`, and `y=20` (since `t` is just the variable of integration, we do not need to give it a value). Move to the `v` line, press **F2** (`Solve`), and we get the same answer as in Figure 18.5. The `left-rt` value that appears is an indication of the answer's accuracy.

Next we solve for an investment rate given a present value of $500,000. Remember that recalculation takes place when you place the cursor on the desired variable's line and press **F2** (`Solve`). We find a present value of half a million corresponds to an investment rate of about 20% ($r = .196...$ in Figure 18.6).

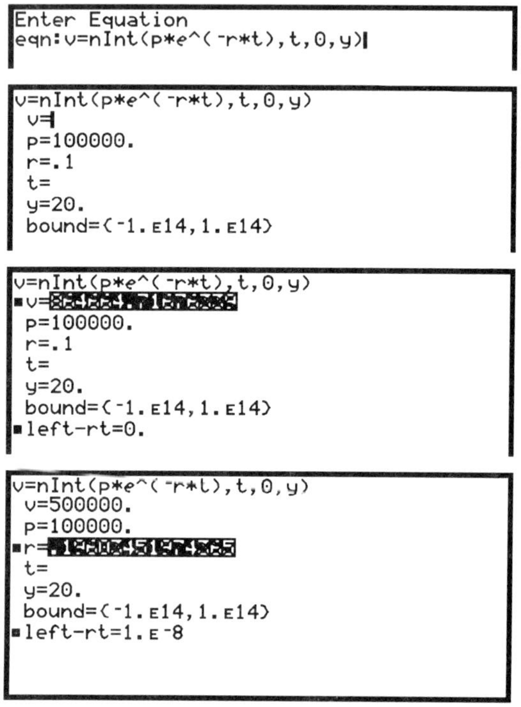

Figure 18.6 `Numeric Solver` *finding the present value V and then an investment rate r. (TI-92 Plus only.)*

Discrete vs. continuous compounding

The analysis above makes the assumption that the income is coming continuously, effectively every second, which is not the case. In real life, the money comes in discrete payments: the first now ($t = 0$), the second in a year ($t = 1$), and so on until the twentieth payment ($t = 19$). If we

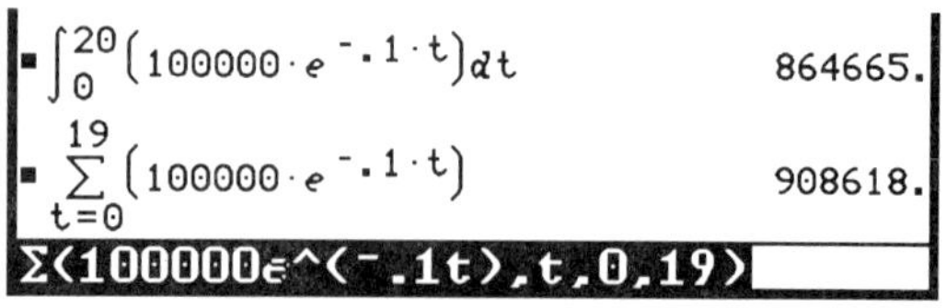

$$\int_0^{20} \left(100000 \cdot e^{-.1 \cdot t}\right) dt \qquad 864665.$$

$$\sum_{t=0}^{19} \left(100000 \cdot e^{-.1 \cdot t}\right) \qquad 908618.$$

Figure 18.7 The present value is greater when discrete payments are made at the start of each year.

want to calculate this to the penny, we need to use a discrete sum for the twenty years. The point is that there is a close relationship between the integral and the sum of a payment sequence. The integral must be used when the income is continuous, but can be used as an approximation for a discrete sum. The sum, about $900,000 (see Figure 18.7), is more than the integral calculation because you receive all $100,000 at the start of the year. You can think of the continuous model as being paid about three-tenths of a cent per second for the next twenty years.

The future value

Since the probability of winning the lottery is essentially zero, you might want to create a jackpot for yourself by investing $100,000 a year (and remember, that's just three-tenths of a cent per second). Your total after T years is called the future value and is given by the definite integral

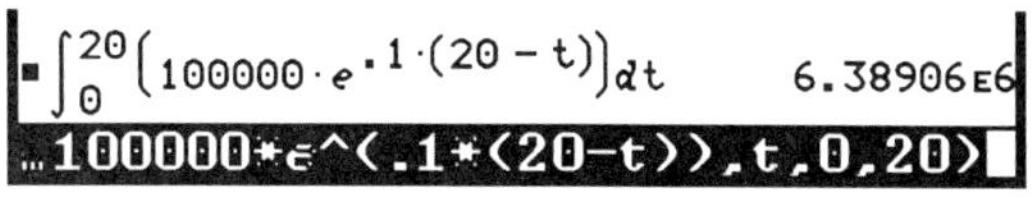

Figure 18.8 The future value in 20 years.

$$V = \int_0^T Pe^{r(T-t)}dt$$

You can find that in twenty years you would have a real jackpot worth over six million dollars as shown in Figure 18.8.

Sometimes the exact form will show a hidden relationship. Change the decimals to fractions in the equations (or specify EXACT mode) and recompute the integrals for present and future value. We find that

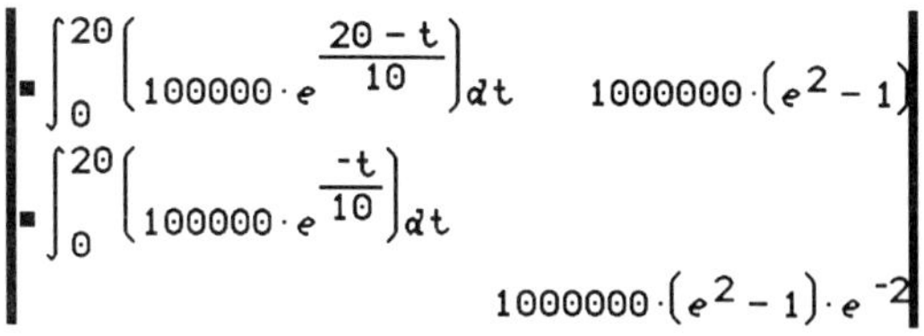

Figure 18.9 The present value and future value in exact form.

$$\text{present value} = \text{future value} \cdot (e^{-2})$$

This is an example of the usefulness of the Fundamental Theorem: you would not be able to account for the e^{-2} factor without writing the integrals in symbolic form.

Modeling: normal distributions

In statistics, a normal distribution has a graph that is a bell-shaped curve. Its general equation is

$$nd(x,\mu,\sigma) = \frac{1}{\sigma\sqrt{2\pi}}e^{-(x-\mu)^2/(2\sigma^2)}$$

where μ is the mean and σ is the standard deviation. We can define this function as nd(x,μ,σ) (because of its two-letter name, it will remain defined even when we clear variables). The Greek letters are entered by using the 2nd_CHAR key (above +, there are shortcuts given in Chapter 16 of the TI-92 *Guidebook* in case you need to enter numerous Greek characters). We then use this to graph the function with specific values of μ = 0 and σ = 1. The resulting curve is called the *standard* normal curve and is shown in Figure 18.10 with $-3 \leq x \leq 3$ and $-.1 \leq y \leq .5$. We saw an example of this family in the last section of Chapter 12.

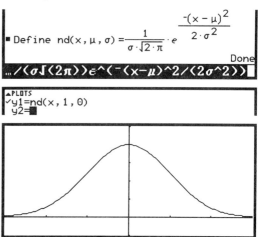

Figure 18.10 The standard normal curve.

The mean and standard deviation

Two fundamental concepts from statistics are the mean of a set of values (casually called the average) and the standard deviation, a measure of how spread out the values are. The commands mean and stdDev, available in the CATALOG, will find these values for any list. If a list has been previously stored and you are not sure of its name, then you can paste it from the 2nd_VAR-LINK menu. For example, in Figure 18.11,

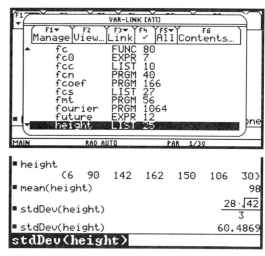

Figure 18.11 Finding the mean and standard deviation for values of a list.

we press H and find the variable height that we used in a previous chapter. Next we find the mean and standard deviation of this list.

The Anchorage annual rainfall

One application of the normal distribution is to model situations where measurements are taken under conditions of randomness. For example, suppose you look at the records for annual rainfall in Anchorage, Alaska over the past 100 years. Let's simplify and say that you found the average of these averages to be 15 inches. Let's say that the standard deviation was 1.

We can estimate the fraction of the years that rainfall is between

(a) 14 and 16 inches, (b) 13 and 17 inches, and (c) 12 and 18 inches

by taking three integrals of the normal distribution with $\mu = 15$ and $\sigma = 1$. Using F5 (Calc), we see from the graphs in Figure 18.11 that the model predicts

(a) 68% of the years have rainfall between 14 and 16 inches,
(b) 95% of the years have rainfall between 13 and 17 inches, and
(c) 99% of the years have rainfall between 12 and 18 inches.

➤ *Tip: After graphing an integral, use F4 (ReGraph) to clear the previous drawing.*

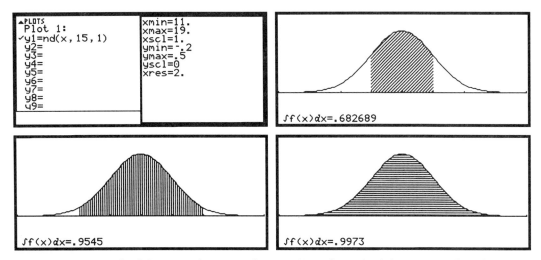

Figure 18.12 Graph of the normal curve with mean 15 and standard deviation 1, then the integrals with limits (a) $14 \leq x \leq 16$, (b) $13 \leq x \leq 17$ and (c) $12 \leq x \leq 18$ that confirm the famous 68-95-99 rule of normal distributions.

A note about TI statistical features

We will use statistics again in Chapter 25, but the TI-92 has powerful statistical features that are not addressed in this book. See Chapter 9 in the TI-92 *Guidebook*.

A. INTEGRATION BY PARTIAL FRACTIONS

Although the TI-92 has many integration techniques built in, you will still want to learn how these methods work. A technique applicable to rational functions uses partial fractions, expressing the ratio of polynomials as the sum of several simpler ratios based on the factorization of the denominator. Without invoking the "sledgehammer" of ∫, we can use the algebra command **expand** to illustrate this technique.

A rational function with distinct linear terms

A common exercise in algebra is to find the common denominator of several terms and combine these into one fraction. In the **F2** (**Algebra**) menu, the command **6:comDenom** does this, as shown in Figure A.1. For purposes of integration, we want to reverse this procedure since it easier to integrate a sum of simpler terms. The command **3:expand** is the "inverse" of **comDenom**.

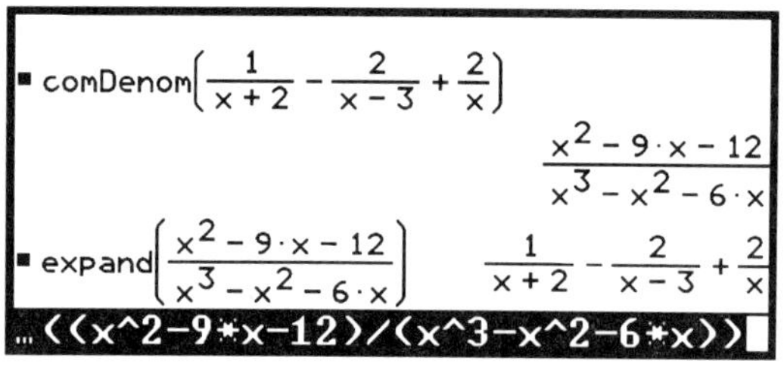

Figure A.1 Algebra commands **comDenom** *and* **expand**.

Integrating the rational function then reduces to three applications of

$$\int \frac{1}{x+a} dx = \ln|x+a| + C.$$

In Figure A.2, notice that the form of the integrand is irrelevant. The TI-92 Plus answers in the traditional format, shown on the left, are compared to the TI-92 without Plus answers which are a single natural log function (the form of the integrand is irrelevant). To see that these answers are equivalent, recall the rules for simplifying and combining logarithms.

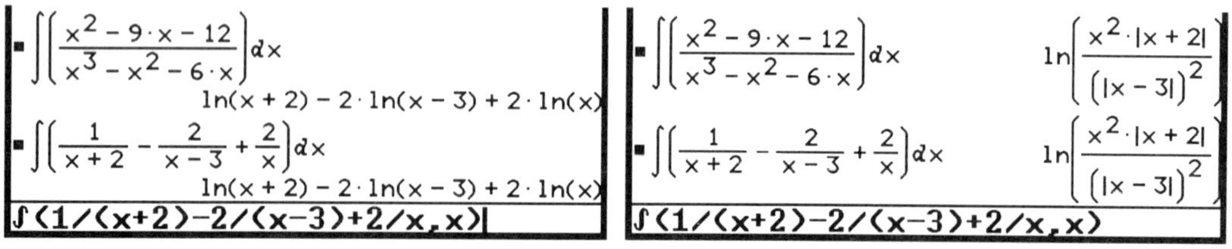

Figure A.2 Integrals of a rational function on the TI-92 Plus and the TI-92 without the chip.

Other rational functions

An initial step in partial fractions integration problems is performing polynomial division so that the degree of the numerator is less than that of the numerator; **expand** does this as well. We can easily integrate power terms and terms where the denominator is linear. When a linear term occurs in the denominator to the second or higher power, there will be additional terms in the expansion. And, of course, not every polynomial factors over the real numbers into linear terms. See Figure A.3.

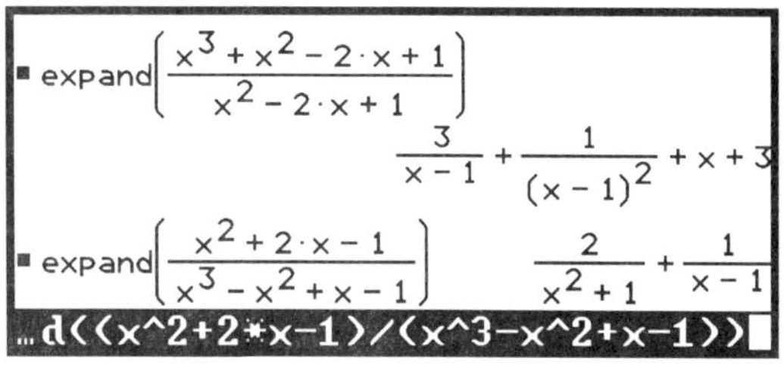

These cases require a few additional integration facts, such as the power rule and some trigonometric substitutions. We conclude with an example showing how using the **expand** command on a rational function will show the terms that have been integrated term by term. In Figure A.4 we first see an integral solution that might be mysterious without understanding the partial fractions technique. Next, **expand** breaks the integrand into three terms which we can match to the three terms of the solution. Finally, as a check, we integrate the first of the three **expand** terms to see the integral of a function with a quadratic denominator.

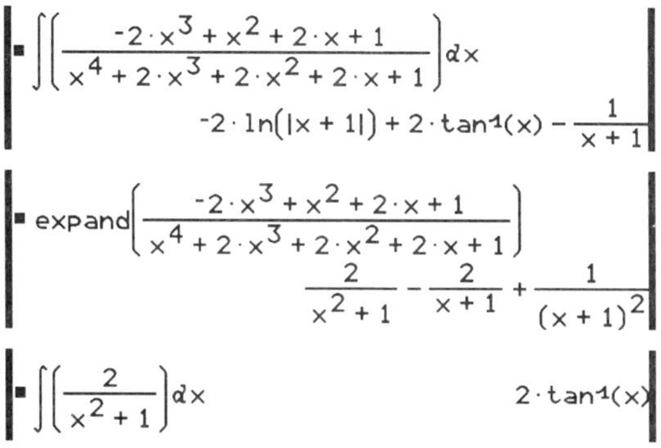

Figure A.3 Expansions of more complicated rational functions.

Figure A.4 A complicated rational function integral elucidated by partial fractions.

B. THE USE AND CONVERSION OF UNITS

Applying calculus to physics has had spectacular success. A notable example is applying integration to find work done. In this demonstration we will show how the TI-92 Plus helps us handle the myriad of units used in physics and other sciences. Also, because there are two main systems of units operating in the world, it is often necessary to make conversions.

The TI-92 Plus allows you to have one of three systems as the default conversion mode. This is selected in the MODE menu (Page 3). An example of the difference is that in SI mode all length will be reported in meters, while in ENG/US the same length would be reported in feet. We will see how we can even have measurement reported in units that we have made up.

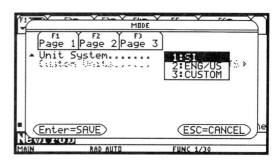

Figure B.1 The Unit System *options in the* MODE *menu (*Page 3*).*

Entering units or constants

The key to understanding the TI-92 Plus system is to think of units (and constants) as multipliers that start with an underline. Although unmarked, the underline symbol is the 2nd version of the P key. With the conversion mode set to SI, if we enter 3_ft, then a meter length is returned. As shown in Figure B.2, the symbol for meters is _m.

Figure B.2 3_ft *is converted to meters.*

How do we know feet are _ft, inches are _in and light-years are _ltyr? A complete menu of units and constants is activated by ♦_P and shown in Figure B.3. Pressing 2nd and an up/down arrow pages through the menu and the right arrow opens submenus. However, these are only the abbreviated names and you may need to consult the "List of Pre-Defined Constants and Units" section of TI-92 Plus *Guidebook* for full names.

The list of constants is partially shown in Figure B.3. Because the constants cover such a range, they are hardest to guess from their symbol. One constant we will use is the acceleration due to gravity, _g.

B. THE USE AND CONVERSION OF UNITS

Figure B.3 Menus of units and constants (♦ _P).

Constants are by default converted to the units of the current conversion mode. For example, in Figure B.4, when the conversion mode is **ENG/US**, then the gravity constant is shown in feet per second squared. But when the mode is **SI**, then the gravity constant is shown in meters per second squared.

Figure B.4 The gravity constant in different systems. Between entries, we changed the **Unit System** *from* **SI** *to* **ENG/US**.

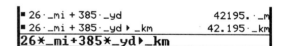

Figure B.5 The length of a marathon (we're back in the **SI** *system).*

Unit conversion

Length

As a simple example, you may want to know how many meters you would run in a marathon. The dictionary tells you that it is 26 miles and 385 yards. We just enter this as an addition in Figure B.5, even with two different units, and the answer is shown in meters. Perhaps you also want the answer in kilometers. That's easy since the **SI** system is based on powers of ten; you don't need a calculator to know that answer. But since we know the answer, this is a good test for our first explicit conversion. To convert a unit we use the convert symbol ▸, the **2nd** version of the **Y** key.

Work

In physics, the unit of work is computed in a different way depending upon the system of units being used. Lifting a five pound book three feet off the floor will result in work expressed in the unit foot-pounds.

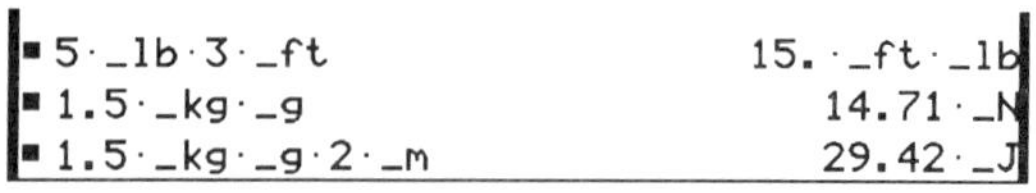

Figure B.6 Work is calculated differently in the two systems (first entry is in ENG/US, the other two SI).

But in the SI system, to find the work in lifting a 1.5 kg book two meters off the floor, we must first convert mass to weight before calculating work. The SI weight (or force) unit of measure is the newton (_N). Work in the SI system is measured in newton-meters. However, these are commonly called joules (_J). All these conversions are shown in Figure B.6.

How much work does it require to pull up a 28-meter uniform chain of mass 20 kg that is dangling from a building? This kind of problem requires the use of integration. We first find the weight of one meter of chain (about 7 newtons) and then find

work done on one small piece ≈ ($7\Delta y$ newtons)(y meters) = $7y \cdot \Delta y$ joules

The total work is given by the integral shown in Figure B.7.

Figure B.7 may lead you to an incorrect conclusion. Usually you cannot use units in integration problems. It was acceptable here only because the units acted as constants that were factored to the outside before integration. It is not

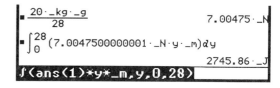

Figure B.7 Finding work done to lift a chain.

allowable to find area under the sine curve by using `sin(x*_m)` as an integrand with units.

➤ *Tip: Applying units to a variable name requires that the name and the underline be separated by either a space or multiplication sign. Otherwise, the whole expression is assumed to be a single variable. When applying units to a number, the space or multiplication sign is optional, but multiplication will be applied to all expressions in the history area.*

Custom units

In case you have come across a reason to use special units, they can be defined and used in calculations. In Chapter 1 we posed the problem of estimating how many sheets of paper are needed for the stack to reach the moon. We can define a new unit of measure called the sheet. A stack of about 300 sheets makes an inch, so our definition is

$$1/300\_in \rightarrow \_sheets$$

Now knowing that the distance to the moon is roughly 230,000 miles, we can find the distance to the moon in our own new units. We just enter

$$230000\_mi \triangleright \_sheets$$

and see the answer in Figure B.8. Recall that the convert symbol is the **2nd** version of Y.

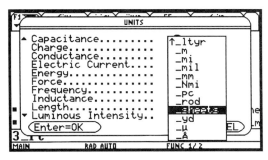

Figure B.8 Creating a user-defined unit.

A helpful feature is that user-defined units are placed in the **UNITS** menu. Figure B.9 shows the newly augmented list.

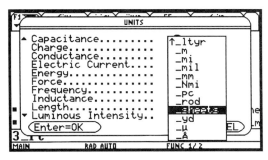

Figure B.9 **Sheets** *appears in the* **UNITS** *menu.*

PART IV
SERIES

19. TAYLOR SERIES AND SERIES CONVERGENCE

20. GEOMETRIC SERIES

21. FOURIER SERIES

DEMONSTRATION: INFINITE SERIES

19. TAYLOR SERIES AND SERIES CONVERGENCE

The tangent line to $f(x)$ at $x = a$ can be considered as the best first-degree polynomial that approximates the function near that point. For most functions, the best quadratic approximation will be a better model. The best degree n polynomial approximating $f(x)$ for x near a, called the Taylor polynomial, is given by

$$P_n(x) = f(a) + f'(a)(x-a) + \frac{f''(a)}{2!}(x-a)^2 + \cdots + \frac{f^{(n)}(a)}{n!}(x-a)^n$$

The TI-92 has a built-in Taylor polynomial generator. Using the sigma sum and higher-order derivative functions, we can also create our own Taylor polynomial function. This is shown in Figure 19.1, where we compare our formula to the built-in function output. (The factorial symbol ! is 2nd_W.) The example is the Taylor polynomial of degree 3 for $y = e^x$ centered at 0.

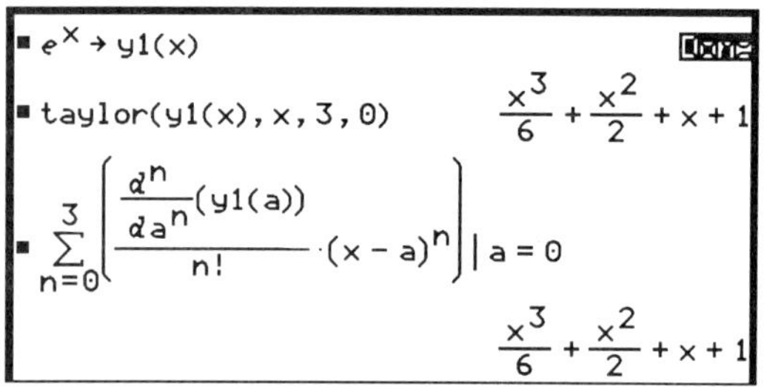

Figure 19.1 Comparing a sigma sum formulation with the built-in Taylor polynomial command.

We can easily verify the Taylor polynomials shown in Figure 19.1 since the function $y = e^x$ is its own derivative and the higher order derivatives are all the same. With a center at $x = 0$, we have $e^0 = 1$ and thus $1/(n!)$ is the coefficient of x^n.

In this chapter we will use the built-in command with syntax

taylor(*function, variable, degree, center*)

to generate Taylor polynomials.

The Taylor polynomials for $y = e^x$

We want to compare the graph of a Taylor polynomial to the original function. For example, we continue $y = e^x$ but find the sixth degree Taylor polynomial with center 1. We can enter the taylor command directly in the Y= editor, shown as y2 in Figure 19.2, but to speed up the graphing we evaluate taylor on the HOME screen and paste the result into y3. We set the graph style of y3 to Path in order to create a little animation.

From the graphs in Figure 19.2 we see that the approximation is quite bad before $x = -1$ and then gives a good fit for the rest of the screen. We can only expect a polynomial to do a good job locally since $y = e^x$ approaches 0 as $x \to -\infty$

19. TAYLOR SERIES AND SERIES CONVERGENCE 137

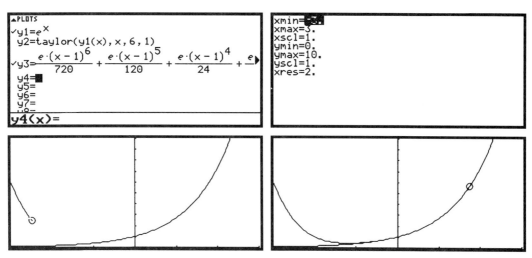

Figure 19.2 *A Taylor polynomial of degree 6 centered at 1 approximates* $y = e^x$ *well for values close to 1.*

and no (nontrivial) polynomial does that. However, this does not mean that for some large negative value of *x* there is no convergence. Given any value of *x*, we can find a degree *n* large enough so the Taylor polynomial centered at 1 provides a close fit. This is not the case for all functions, as we see in the next example where the convergence is limited to a finite interval.

The Taylor series for ln(*x*)

As a second example, we find the Taylor polynomials of degree 5, centered at 1 for the function ln(*x*). Again, the formula is pasted into y3. The Taylor polynomial graph is shown in Figure 19.3 with a **Thick** graph style. It can be seen that it is not very good for *x* > 2 and has defined values for *x* ≤ 0 where the original function is not defined. However, near *x* = 1 it is an excellent approximation. In the case of e^x, we claimed that for any *x*-value there was an *n* for which the Taylor polynomial would convergence to the function. For ln(*x*) and the Taylor approximations centered at 1, no matter how large *n* is, the polynomial will never be close to ln(*x*) for x ≥ 2 (The ratio test gives a radius of convergence of 1 about *x* = 1.)

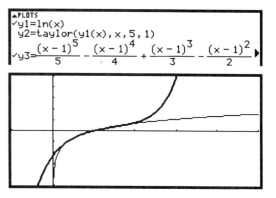

Figure 19.3 *The Taylor approximation of order 5 centered at 1 calculated and pasted to* y3. *The graph window is -1 ≤ x ≤ 5 and -5 ≤ y ≤ 5.*

The Taylor series of the sine function

In the next example, we will increase the degree and see the Taylor polynomials centered at 0 wrap closer and closer to $y = \sin(x)$. In Figure 19.4, only the graphs for degrees 1, 3, 5, 7 are shown since the graphs for degrees 2, 4, 6, 8 (respectively) are the same. This is because the even power Taylor coefficients are zero. Looking at the four graphs, we can see that the approximations are close on wider and wider intervals as the degree gets larger. The interval of convergence for the seventh degree Taylor polynomial seems to be $-\pi < x < \pi$, but increasing the degree will show an increasingly wider convergence. The ratio test can be used to confirm that the interval of convergence for these Taylor polynomials expands to include all reals.

➤ *Tip: The higher the degree of the polynomial, the longer it takes to graph. You may want to set* `xres=4`.

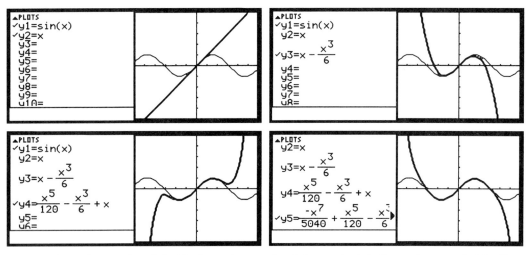

Figure 19.4 *The graphs of the Taylor polynomials of degree 1, 3, 5, 7 are shown in bold. The window is* `ZoomDec` *in the split screen.*

These approximations of trigonometric and exponential/logarithmic functions are examples of the fact that any infinitely differentiable function can be locally approximated by polynomials. This is important because computers and calculators are masters at polynomial evaluation — after all, it is just addition and multiplication.

Evaluating a Taylor polynomial at $x = 1$ is the same as summing the coefficients. As the degree of the Taylor polynomial goes to infinity, the value at $x = 1$ remains finite because of the $n!$ term in the denominator of the Taylor coefficients. Another way of saying this is that the series of Taylor coefficients converges. This leads to a more general question.

How can we know if a series converges?

This is a difficult question, but we have the ratio test and the alternating series test to help us in many cases. If asked for proof, you will need to use an analytical argument, but a graph or table will often tell you what whether to pursue a proof of convergence or divergence. Since you are investigating a series, the n-th term is given symbolically (or else you will need to write it symbolically). In the example below, the sigma summation notation is given, but if a series is listed without a symbolic term, then the first order of business is to write it as a sigma sum. We will use y1 = Σ(...) to either graph or make a table of values from which we form an opinion about whether the series converges or diverges.

The harmonic series: $1 + \frac{1}{2} + \frac{1}{3} + \frac{1}{4} + \ldots = \sum_{n=1}^{\infty} \frac{1}{n}$

In Figure 19.5 we first enter this infinite series as

$$\Sigma(1/n, n, 1, \infty)$$

and find that we get no information from this approach. Next we translate the partial sums of the harmonic series into a function that sums the first through x-th terms,

$$y1 = \Sigma(1/n, n, 1, x)$$

and graph with xmin=0 and xmax=300. The graphing gets progressively slower as x gets larger; remember that you can press ON to interrupt at any time. The partial sums appear to be continually increasing without limit. This means the series does not converge, but analytic means are needed to be absolutely sure. (You can use a comparison with $\int_{1}^{\infty} \frac{1}{x} dx$.)

You might ask why don't we just find the 10,000th partial sum from the HOME screen (since the limit value would likely be clear by then). This can be done and the answer will be just over 12, but you will pay dearly

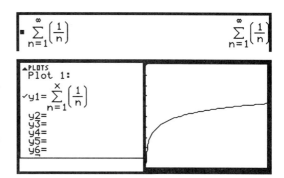

Figure 19.5 Suspecting divergence from a partial sum graph with $0 \leq x \leq 300$ and $0 \leq y \leq 10$.

in time and battery use. This calculation takes about one hour on a TI-92 Plus! Also, without other values for comparison, it is unclear from just this one whether the series is converging or diverging.

The alternating harmonic series: $1 - \frac{1}{2} + \frac{1}{3} - \frac{1}{4} + \ldots = \sum_{n=1}^{\infty} (-1)^{n-1} \cdot \frac{1}{n}$

Often series alternate positive and negative terms. This sign switch is cleverly written in sigma notation using a power of negative one. Since every other term is subtracted, we expect the range of the partial sums will not be as great as for the harmonic series. Any alternating series will converge if its terms approach zero as *n* becomes infinite.

When we enter an infinite series on the TI-92 to find a finite sum, we may be greeted with a transformation of the series and no limiting value. The conversion to `cos(n·π)` in Figure 19.6 is yet another clever way of writing a quantity that will alternate between 1 and -1.

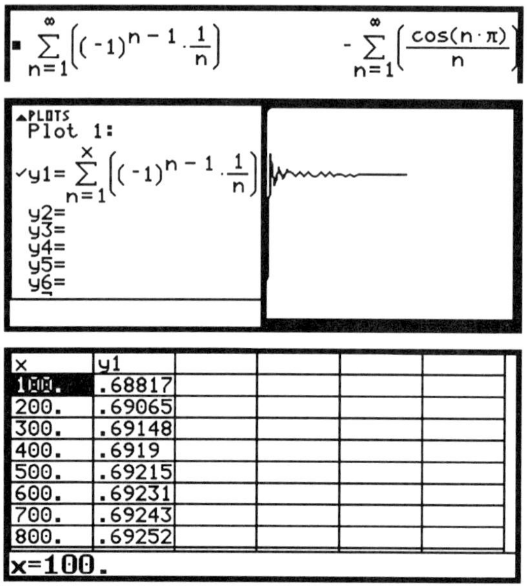

We define the partial sums function

$$y1 = \Sigma((-1)^{\wedge}(n-1)*(1/n), n, 1, x)$$

Figure 19.6 Suspecting a convergence, we break before the graph is completed. The window is $0 \leq x \leq 300$ and $0 \leq y \leq 1$. We can also see evidence for convergence in the table values.

and graph it. Press **ON** to break when you are satisfied. We see from the graph in Figure 19.6 that it is a good bet that this series converges. But unless we waited for the complete graph and used **Trace**, we would not have a very accurate convergence value from the graph. Instead, we use a table to find a good limiting value.

▶ *Tip: When writing a sigma sum symbolically, it is easy to make small errors. Before graphing or making a table, you might want to make a small list of terms to verify that your notation is correct.*

▶ *Tip: If you do not want a sum to be a rational number answer, enter the start or stop value as a decimal to put the answer in the* **APPROXIMATE** *format.*

A fast converging series: $2 - \frac{2}{3} + \frac{2}{9} - \frac{2}{27} + \ldots = \sum_{n=0}^{\infty} (-1)^n \cdot \frac{2}{3^n}$

The terms of this series approach zero as n gets large, so by the alternating series test it converges. Looking at a table, we see in Figure 19.7 that by the 15th term several digits of accuracy are assured and we can assume (correctly) that the exact sum is 1.5.

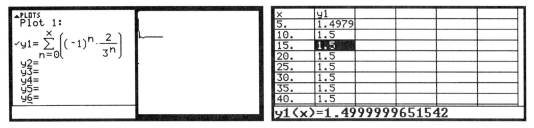

Figure 19.7 Making a graph and table of a fast converging series.

A slow converging series: $1 - \frac{1}{3} + \frac{1}{5} - \frac{1}{7} + \ldots = \sum_{n=1}^{\infty} (-1)^{n-1} \cdot \frac{1}{2n-1}$

Unlike the last example, some series converge very slowly. You may know a series is convergent, but find getting a highly accurate value can be difficult. In Figure 19.8 we see a graph that looks convergent but the table shows that even after 800 terms we can only comfortably predict that the value is perhaps between 0.78 and 0.79. It turns out that that the true value is $\pi/4 = 0.785398\ldots$ This amazing fact can be shown by finding the Taylor series for arctan(x) and recalling that arctan(1) = $\pi/4$.

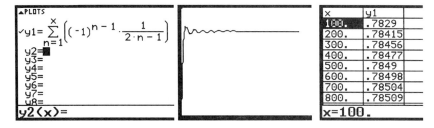

Figure 19.8 A slow converging graph and table.

In general alternating series are not evaluated with exact limits by the TI-92. This is why we have shown the numerical and graphical approaches in this section. We were lucky to be able to gave exact limits in the last two examples but in general the limit can be difficult to find in exact form. See Demonstration A on Infinite Series at the end of this Part.

20. GEOMETRIC SERIES

In the last chapter, we found Taylor polynomials by starting with a function and forming the successive coefficients for powers to find a series representation. Sometimes we face the reverse situation: the series is known but the function is not. We now consider finding the function from a given series. In some cases, the TI-92 can write a general formula for a sum.

The general formula for a finite geometric series

A finite geometric series has the form

$$a + ax + ax^2 + \ldots + ax^{n-1} + ax^n$$

Note that the coefficient a is the same for each term. The closed form sum is

$$S_n = \sum_{i=1}^{n} ax^{i-1} = a\frac{(1-x^n)}{(1-x)}, \quad (x \neq 1)$$

The TI-92 will verify this finite sum formula, but the output will be two fractional expressions. In Figure 20.1, we show two F2 (Algebra) commands that transform the sum formula closer to the traditional.

You might think that such a sum would be so rare as to make it not worth considering, but this kind of finite series comes up in many situations.

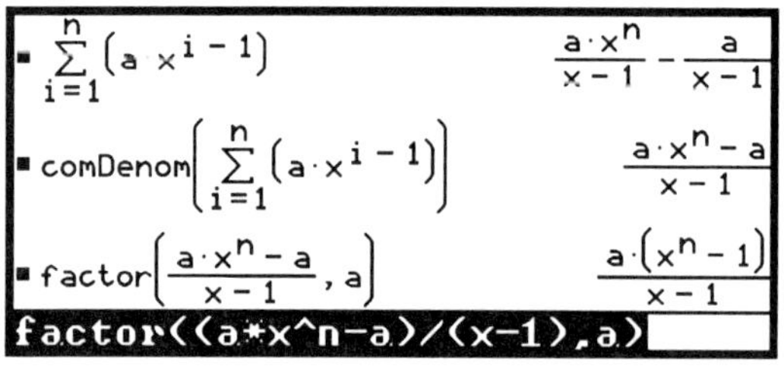

Figure 20.1 The finite geometric series sum.

Repeated drug dosage using Σ

Consider a 250 milligram dose of an antibiotic taken every six hours for many days. The body retains only 4% of the drug present in the body after six hours. Be careful: this does not say that 4% of 250 mg is left. This is only the case at the end of the first six hours. The interesting part is that at the end of the second six hours, the body retains 4% of the second dose *and* 4% of the remaining first dose. Let's make a sequence of the amount of drug in the body right after taking the n-th dose.

$$Q_1 = 250,$$

$$Q_2 = Q_1(0.04) + 250,$$

$$Q_3 = Q_2(0.04) + 250, \text{ etc.}$$

But if we substitute lower sums and multiply, we find

$$Q_2 = 250(0.04) + 250 \text{ and}$$

$$Q_3 = 250(0.04)^2 + 250(0.04) + 250$$

In general,

$$Q_n = 250(0.04)^{n-1} + \ldots + 250(0.04)^2 + 250(0.04) + 250$$

The Q's are a finite geometric series with $a = 250$ and $x = 0.04$. Let's calculate the amount of antibiotic in the body at the time of taking the fourth pill. In Figure 20.2 we use Σ to find the total amount left after the fourth pill. This is the same answer we get from the series formula.

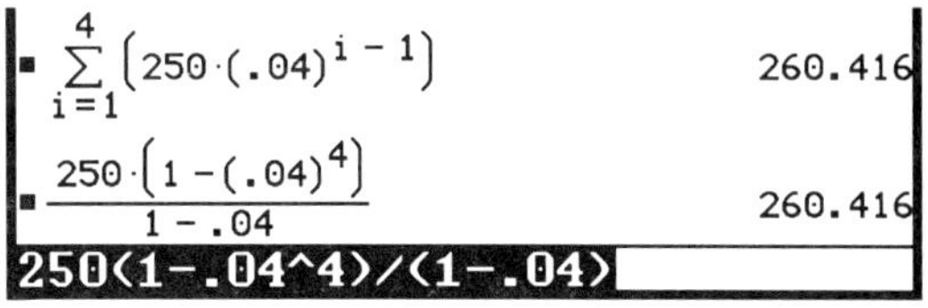

Figure 20.2 The dosage present in the body at the fourth dose summed as a sequence and compared to the formula.

Using a table to find consecutive finite sums of a sequence

The problem with the above calculation is that it is isolated; we have no idea about how it is changing as *n* increases. To see consecutive finite sums of a sequence, you can define a function using Σ and set up a table to show the desired values. Clearly, the table setting must make the *x*-values positive integers. This is shown in Figure 20.3, where we also use ♦_F to increase the cell width to 12. Notice how the amount of medication stabilizes within the first day (by the fifth dose).

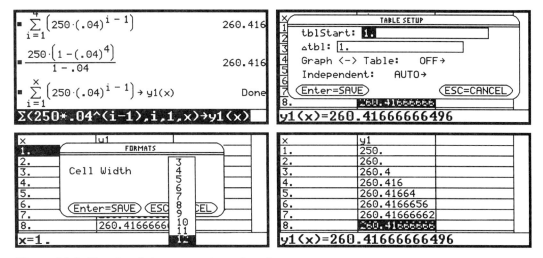

Figure 20.3 Showing finite sums using a function.

A different drug might have a much higher retention level. Defining y2 from y1 with 0.50 instead of 0.04 gives the table in Figure 20.4. A drug with this retention level will take about two days to reach its stable limit, near 500 mg.

Figure 20.4 Retention amounts for the rates 4% and 50%.

Regular deposits to a savings account

Another direct application of the finite geometric series is finding the value of an investment that earns a periodic interest. Suppose that you plan for retirement by putting $1000 a year into a savings account that earns 5% annual interest. You might want to know how much this will be worth after the n-th deposit. We find this in Figure 20.5 using two methods. First, we enter the summation into y1 and make a table showing the value every ten years after the first year. Second, we just use the finite sum formula to check the sum when $n = 41$, when you might be considering retirement.

Figure 20.5 The value of a 5% savings account starting with $1000 after 41 annual deposits.

Identifying the parameters of an infinite geometric series

Evaluating a series at some point yields a numeric sum and we want to know if it will converge or diverge. If the numeric series is geometric, then we can find the sum quite easily. We are ready to practice on the following list of infinite series.

(a) $1 + \dfrac{1}{2} + \dfrac{1}{4} + \dfrac{1}{8} + \cdots$

(b) $1 + 2 + 4 + 8 + \cdots$

(c) $6 - 2 + \dfrac{2}{3} - \dfrac{2}{9} + \dfrac{2}{27} - \cdots$

We want to see if they are geometric sums like $a + ax + ax^2 + \cdots + ax^n \cdots$, so we need to identify a and x: this is the hardest part. (If the sum is given in sigma sum notation, then this task is essentially done for us.) In the above cases, we see that each series is a geometric series with a as the first term and we can find x from the second term. The TI-92 will evaluate the first two sums as shown in Figure 20.6. However, it will not evaluate the third sum analytically and we have to use a table to find its convergent value.

We investigate all three series by defining the partial sum functions in `y1` to `y3`. Remember that `Y=` definitions require `x` as the independent variable, so we replace n with x even though x is usually the number being exponentiated. By checking the table, you can quickly find that series (a) converges to 2, (b) diverges, and (c) converges to 4.5.

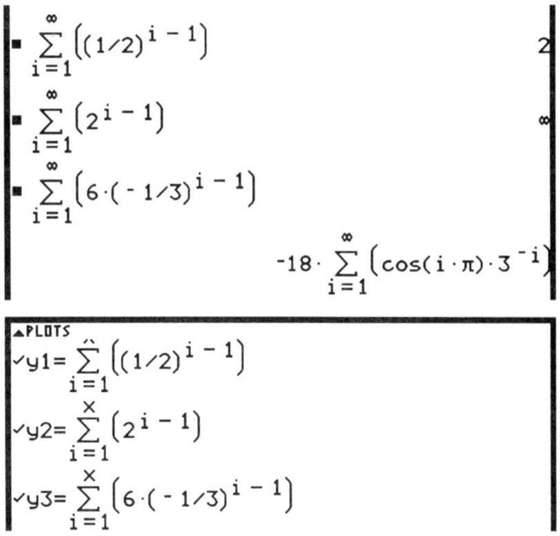

x	y1	y2	y3
1.	1.	1.	6.
11.	1.9990234	2047.	4.5000254
21.	1.999999	2097151.	4.5
31.	2.	2.14748E9	4.5
41.	2.	2.199E12	4.5
51.	2.	2.2518E15	4.5
61.	2.	2.3058E18	4.5
71.	2.	2.3612E21	4.5

x=1.

Figure 20.6 Checking different series from a general definition.

Summing an infinite series by the formula

It should not be surprising that if the value of x is one or greater, then the infinite geometric series will diverge (if $x > 1$, then successive terms are increasing, as with $x = 2$ above). If we restrict x to $0 < x < 1$, then the series will have a finite sum (as with $x = 1/2$ above). In Figure 20.7 we see that a general infinite series will not be evaluated and restrictions which include positive x result in symbolic formulas. But the TI-92 does not evaluate the more general formula that allows x to be negative. This is why the top panel of Figure 20.6 evaluated the first two series but not the third (which had $x = -1/3$). The more general formula is

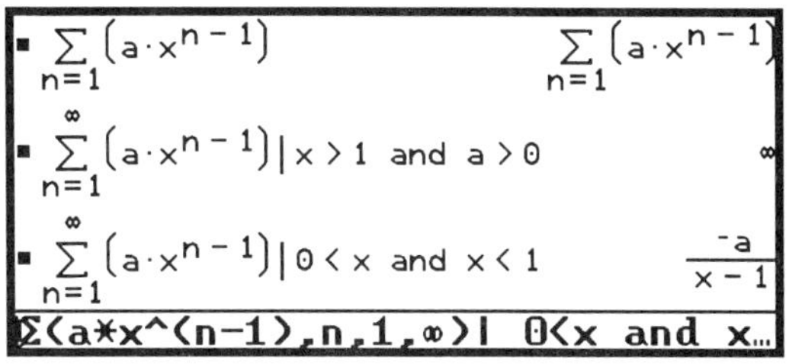

Figure 20.7 Restrictions that give sum formulas for infinite geometric series.

$$a + ax + ax^2 + \ldots + ax^{n-1} + ax^n + \ldots = \frac{a}{1-x}, \quad \text{for } |x| < 1$$

In Figure 20.8 we calculate and compare the formula values with our previous two results and then revisit the formula for the two drug dosage cases. Note that they both qualify as finite sums since percentages (here $x = 0.04$ and $x = 0.5$) satisfy the condition $|x| < 1$.

Figure 20.8 Infinite sums from previous examples by using the formula.

Piggy-bank vs. trust

Suppose that your parents are trying to decide on a plan to provide for your future and they have two choices:

I. Each year they put your age in dollars into a piggy bank.

II. Each year they put $3 into a trust account that earns 6% annual interest.

We already know how plan II works: use $a = 3$ and $x = 1.06$ in the formula above for a finite geometric series. In plan I we need to look at the sum of the series $1 + 2 + 3 + \ldots + n$. This is *not* a geometric series. A clever way to consider this sum is to write it twice, once forward and once backward:

$$
\begin{array}{ccccccccc}
1 & + & 2 & + & 3 & + \ldots + & n-1 & + & n & = & S_n \\
n & + & n-1 & + & n-2 & + \ldots + & 2 & + & 1 & = & S_n \\
\hline
n+1 & + & n+1 & + & n+1 & + \ldots + & n+1 & + & n+1 & = & 2S_n
\end{array}
$$

There are n of these $(n+1)$ sums, so we have $S_n = n(n+1)/2$. We see that the TI-92 is this clever in Figure 20.8. Comparing table values shows that the piggy bank is probably the better plan for you now and for a while. (Notice the changing speed in the calculation of table entries.)

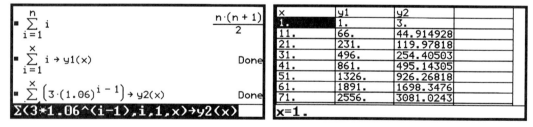

Figure 20.8 The TI-92 knows certain sums. Sigma summations can be used to define functions. The table values show y2 exceeds y1 after 61 and before 71.

To compare graphically, we could graph the two sigma summations as they are now defined, but the graphing would be painfully slow, especially towards the right of the screen (the TI-92 does not build from previous work but recomputes for each *x*-value). Instead, we use the two sum formulas. In Figure 20.9 we show a screen for a nice long life of 100 years. Finding the intersection, we see that plan II is preferable from age 64 and, as you get into your seventies, plan II is much better.

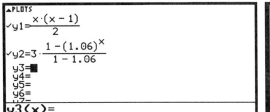
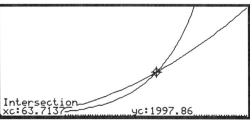

Figure 20.9 Sigma summations are OK for a table, but we switch to the formulas to graph (the window is $0 \leq x \leq 100$, $0 \leq y \leq 5000$). The x-value of the intersection is not valid for the original summations, so we round up to 64.

21. FOURIER SERIES

The Taylor polynomials are good approximations to a function inside the radius of convergence; beyond that, they're awful. Even cases where the radius of convergence is all real numbers can be difficult because the degree needed to force convergence can be astronomical. As suggested by the example of the sine function in Chapter 19, the Taylor polynomials are practically useless for periodic functions because *n* needs to be so large. We now find a set of functions that provide good global approximations of a periodic function. These are called Fourier approximations.

First we will discuss how to create user-defined functions. This is the best way to define most periodic and piecewise defined functions.

A word about user-defined functions

A defined function can be used with the ease of numbers in formulas, graphs and tables. So far we have only defined functions from the HOME screen entry line as a single formula. A more powerful kind of function is available. These user-defined functions can use logical expressions similar to those used in writing a program. But while a program can only be used on the entry line by itself, these more complicated functions can be used within expressions and in the Y= editor.

The basic form of any function is that it uses a variable name gives an output based on one or more independent variables. See Figure 21.1 for a sequence of screens to define a function called `first` that will return a value of -1 if the input is negative and a value of 1 otherwise.

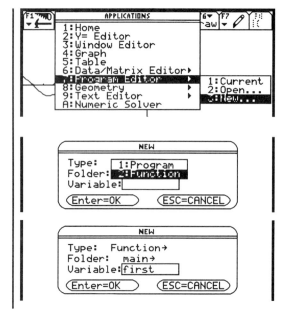

We construct a function like we write a program. Start with the sequence APPS 7:Program Editor 3:New. You can also 2:Open and select from the list of currently defined functions.

At the top of the dialog box, change Type to 2:Function.

Give the function the variable name `first`. Press ENTER twice to get to the edit screen.

21. FOURIER SERIES   149

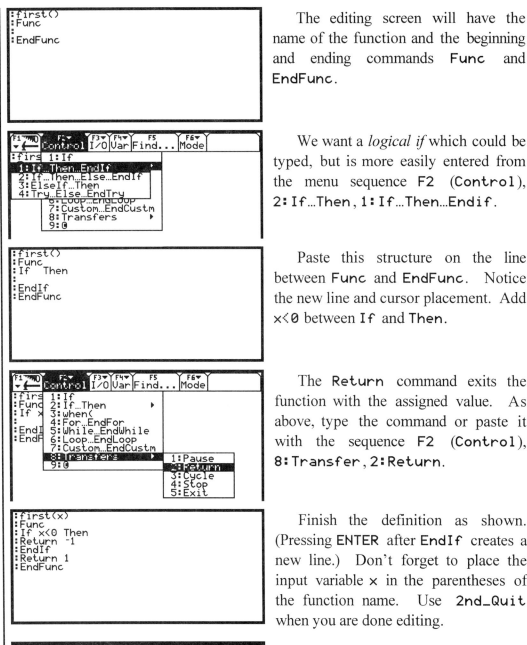

The editing screen will have the name of the function and the beginning and ending commands `Func` and `EndFunc`.

We want a *logical if* which could be typed, but is more easily entered from the menu sequence F2 (`Control`), 2:`If…Then`, 1:`If…Then…Endif`.

Paste this structure on the line between `Func` and `EndFunc`. Notice the new line and cursor placement. Add x<0 between `If` and `Then`.

The `Return` command exits the function with the assigned value. As above, type the command or paste it with the sequence F2 (`Control`), 8:`Transfer`, 2:`Return`.

Finish the definition as shown. (Pressing ENTER after `EndIf` creates a new line.) Don't forget to place the input variable x in the parentheses of the function name. Use `2nd_Quit` when you are done editing.

Graph the `first` function to show that it is working properly.

Figure 21.1 The sequence of screens to define and graph a function `first`.

User-defined functions

The only kinds of periodic functions we have considered so far are trigonometric. We now introduce and graph a few other types.

The square wave function

This function is commonly used in electrical engineering to model switching: it is either on or off. We use 1 to mean on, 0 to mean off. In this case we define the function again using the logical structure If...Then...EndIf.

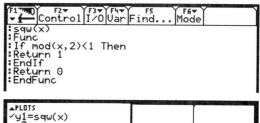

The complete definition is shown in Figure 21.2. The mod function is built into the TI-92; mod(x,2) returns the remainder after x has been divided by 2. Note that all of its values must be between $0 \leq y < 2$. We graph the function, sqw(x) with the Dot style.

Later in the chapter we will determine and graph Fourier approximations to this function.

*Figure 21.2 A logical **If** expression in a user-defined function definition. The window setting is $-3 \leq x \leq 3$, $-1 \leq y \leq 2$.*

Piecewise defined functions

The If...Then structure can be extended to more complicated definitions of piecewise defined functions. For example, we next define pwd(x) (for piecewise defined function) to have three different formulas used for different values of *x*. Figure 21.3 shows its definition and graph. Notice that a piecewise defined function can be continuous or discontinuous. To create even more complicated piecewise defined functions on intervals such as $1 < x < 2$, use 1<x and x<2.

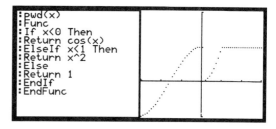

*Figure 21.3 A logical **If...ElseIf...EndIf** structure for a function with three formulas.*

➤ *Tip: If you're editing an existing function with the split screens as above, trying to graph it gives an error message* Variable in use so reference or changes are not allowed *(and it doesn't graph).*

The triangle wave function

Our final example is a triangle wave function which starts at 0, increases linearly to one at $x = 1/2$, then decreases linearly to zero again at $x = 1$. The two pieces of this function can be added together to make a single function definition as shown in Figure 21.4. The line `Local i` allows the variable `i` to be used internally and then returned to its original state after the function is evaluated.

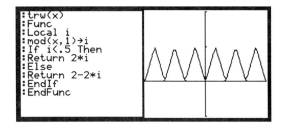

Figure 21.4 A triangle wave function (in *Line* style).

The general formula for the Fourier approximation function

A Fourier approximation function uses the sine and cosine functions to approximate periodic functions. These functions are not polynomials, but we use the polynomial vocabulary to describe them. We use the term *degree* to specify which sine/cosine terms are included and specific constant multipliers are called the *coefficients*. The definition of the *n*-th degree Fourier function for the interval $-\pi \leq x \leq \pi$ is

$$F_n(x) = a_0 + a_1 \cos(x) + a_2 \cos(2x) + \ldots + a_n \cos(nx)$$
$$+ b_1 \sin(x) + b_2 \sin(2x) + \ldots + b_n \sin(nx)$$

With this basic structure in mind, we give the generalized definition which includes the general period *b* and the definitions of the coefficients.

$$F_n(x) = a_0 + a_1 \cos((\tfrac{2\pi}{b})x) + a_2 \cos(2(\tfrac{2\pi}{b})x) + \ldots + a_n \cos(n(\tfrac{2\pi}{b})x)$$
$$+ b_1 \sin(x(\tfrac{2\pi}{b})) + b_2 \sin(2(\tfrac{2\pi}{b})x) + \ldots + b_n \sin(n(\tfrac{2\pi}{b})x)$$

$$a_0 = \frac{1}{b}\int_{-b/2}^{b/2} f(x)\,dx$$

$$a_k = \frac{2}{b}\int_{-b/2}^{b/2} f(x)\cos(k(\tfrac{2\pi}{b})x)\,dx \quad \text{for } k > 0$$

$$b_k = \frac{2}{b}\int_{-b/2}^{b/2} f(x)\sin(k(\tfrac{2\pi}{b})x)\,dx \quad \text{for } k > 0$$

Since the graphs we will create are for periodic functions, we will integrate instead from 0 to *b*. In cases where the Fourier function is approximating a non-periodic function over an interval of length *b*, it is important to have the integration centered on the interval.

A system for graphing the Fourier approximation function

We now automate the creation of the Fourier approximation functions. We will create a simple program `fcoef()` to calculate the coefficients and then define a Fourier approximation function called `fap` for graphing or evaluating values.

Setting up before graphing

Enter the function (probably periodic) to approximate in `y1`. Find a nice window for `y1` that shows several periods on each side of the origin.

Creating the coefficients

We store the coefficients in two lists named `fcc` (Fourier coefficient cosine) and `fcs` (Fourier coefficient sine) and the initial value (a_0) in `fc0`. This is most easily done by typing the commands into a program where they can be stored and later reused. The inputs for the program `fcoef` below are the order and period length; it has no visible output, but creates the lists. The function `fap(n,b,x)` is a direct translation of the Fourier approximation function of degree n, so we can use it to define a graph via the Y= editor. The listings are shown in Figure 21.5.

```
:fcoef(n,b)
:Prgm
:DelVar fcc,fcs
:Local k
:For k,1,n
:2/b*∫(y1(x)*cos(k*(2*π/b)*x),x,0,b)→fcc[k]
:2/b*∫(y1(x)*sin(k*(2*π/b)*x),x,0,b)→fcs[k]
:EndFor
:1/b*∫(y1(x),x,0,b)→fc0
:EndPrgm
```

```
:fap(n,b,x)
:Func
:fc0+Σ(fcc[i]*cos(i*(2*π/b)*x)+fcs[i]*sin(i*(2*π/b)*x),i,1,n)
:EndFunc
```

Figure 21.5 The program `fcoef()` creates the Fourier coefficients which are used in the user-defined Fourier approximation function `fap`.

➤ *Tip: If you create a function on the entry line with `Define` and open it in the `Program Editor`, then you will see no `Func` and `EndFunc` delimiters.*

The Fourier approximation to the rise-and-crash function

To test all these definitions, we create and graph an approximating function for the rise-and-crash function. This is a dramatic name for $x - \text{int}(x)$, similar to `fpart` that we saw in Chapter 9. At each integer input, it starts at zero and rises linearly to one then falls back to zero when it reaches the next integer. Rather than use $x - \text{int}(x)$ when running `fcoef`, we can simplify things. The rise-and-crash function has period 1 and the program only uses one period, so we use x for `y1`: it agrees with $x - \text{int}(x)$ on $0 \leq x \leq 1$ and is much easier for the calculator. In Figure 21.6 you see that using a function that is integrable by a formula leads to exact coefficients. Recall that $\text{int}(x)$ is beyond the TI-92's formula base; running `fcoef`

21. FOURIER SERIES   153

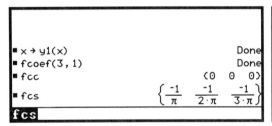 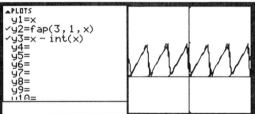

Figure 21.6 Using **fcoef(3,1)** *to generate the coefficients of order 3, with a period 1 for the function y = x. Then a graph of* **fap** *and the actual rise-and-crash function.*

with $x - \text{int}(x)$ raises the `Questionable accuracy` flag and gives decimal approximations for the coefficients. Graphing the Fourier approximation with the periodic function shows a good fit.

The triangle wave function

Next we find an approximation of the triangle wave function as defined in Figure 21.4. It has a period of 1 and the polynomials are a close fit, even with degree 2. The sine curve shape is closer to the triangle wave than the square wave, so it makes sense that the fit is very good even at degree 2. The wait for **fcoef(2,1)** is about one minute and you see that the coefficients are essentially zero for three of the four coefficients. The integration has been done numerically so it is both slow and an approximation.

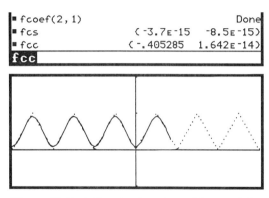

Figure 21.7 A Fourier function of degree 2 fits the triangle wave function very closely.

➤ *Tip: Numeric approximations of coefficients which are zero often appear as non-zero values with negative exponents of -14 and -15.*

The square wave approximations

To approximate the square wave function, we must test our patience with the limited calculation speed of the TI-92 it takes this example takes over 20 minutes to calculate the coefficients. In Figure 21.8, we generate the coefficients only once for

■ fcoef(7,2)   Done
■ fcc
  (3.25E-15   -7.5E-16   9.5E-16   -4.2637▶
■ fcs
  (.63662   1.7E-15   .212207   -1.095E-14▶

Figure 21.8 Fourier coefficients calculated (with **y1=sqw(x)***) for seventh order approximations of period length two. Warning: calculation takes 20 minutes (on a TI-92 Plus).*

order 7. Once the coefficients are generated for the highest power, the lower order polynomials will use the appropriate subsets of the lists. In this case, all values in `fcc` are essentially zero, meaning it does not use the cosine function to approximate this function. Only the `fcs` odd entries contribute non-zero values, so, for example, increasing the order from 3 to 4 results in the same Fourier polynomial.

We see in Figure 21.9 that the fit gets better as the order increases.

➤ *Tip: This calculation time can be reduced by changing `fcoef` to not find `fcc` (replace the integration in that line with `0`).*

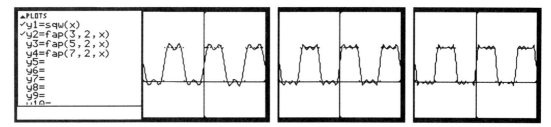

Figure 21.9 Fitting the square wave function with period of length 2, by using Fourier approximations of orders 3, 5 and 7.

A. INFINITE SERIES

We are indebted to our colleague Ahmad Mirbagheri for suggesting this demonstration. Here we illustrate two techniques for investigating series when the straight-forward methods are impossible or impractical.

The comparison test for convergence of a series

Does the following infinite sum converge or diverge?

$$\sum_{n=1}^{\infty} \frac{\ln(n)^2}{n^2}$$

In Figure A.1 we see that the TI-92 gives no immediate answer and trying to sum just 100 terms take so long that we interrupt the calculation. (Use **ON** to break.) Rather than using time-consuming discrete sums, let's try a comparison to the continuous version of this function. Since the summation terms decrease as n increases, the continuous function is always slightly more than the discrete sum in this case. For example, the first term of the discrete sum is zero but the continuous function on $1 \leq x \leq 2$ has a positive integral. We see that the integral of the continuous function converges, therefore we are certain that the discrete sum converges.

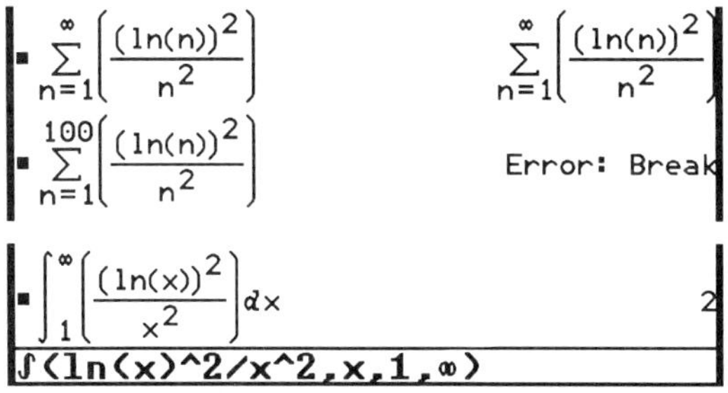

Figure A.1 Attempts at the discrete sum and the integral of the continuous version.

So we know the infinite series converges, but not the exact sum. The next best thing we can do is find an n for which the approximating partial sum is accurate within a given tolerance. Suppose we want the sum to be accurate to within 0.05. We had such success with the continuous integral, let's use it again to bound the error. We use s for the exact sum, s_n for the n-th partial sum.

$$s - s_n \leq \int_{n}^{\infty} \frac{\ln(x)^2}{x^2} dx$$

We see in Figure A.2 that the integral has an exact solution. We use this answer in **solve** to obtain the result that if $n > 1373$ then the sum

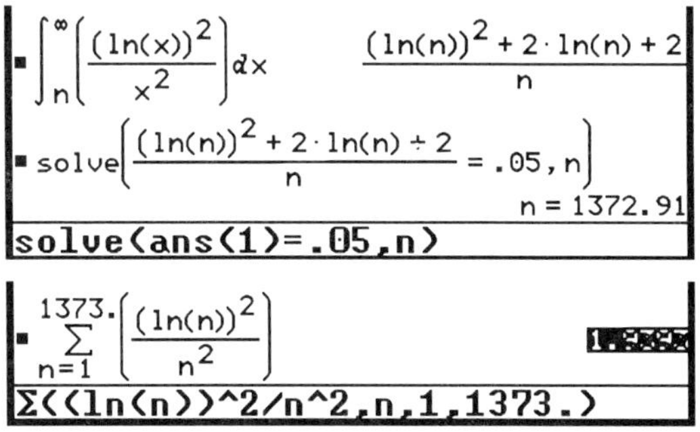

Figure A.2 Determining n so that the partial sum is within 0.05 of the exact sum.

156 PART IV / DEMONSTRATION

will have the desired accuracy. Trying to compute the exact sum with $n = 1373$ would exceed the memory of this calculator, but entering the limit as a decimal puts the calculator into approximate mode and in a mere two minutes we have an answer. We conclude that $1.8893 < s < 1.9893$, which confirms that the integral of the continuous version is greater than the infinite sum.

Seeing convergence from a graph

Consider the partial sums

$$s_1 = 1 + x$$
$$s_2 = 1 + x + x^2$$
$$s_3 = 1 + x + x^2 + x^3$$
$$s_4 = 1 + x + x^2 + x^3 + x^4$$

Graphing these shows a pattern: they are wrapping themselves closer and closer to the function $1/(1-x)$ for the interval $-1 < x < 1$. The function y5 is drawn with the Square style so that it will be bold and not connect across the discontinuity at $x = 1$. Each of the s_n graphs has been labeled with the F7 Text tool. It should be visually clear in Figure A.3 that there will be no convergence for values of $x \geq 1$ nor for $x \leq -1$, no matter how big we make n. You may question whether there will be convergence for an x value such as 0.99, but the pattern shows that as n increases, the s_n pull up to the vertical asymptote.

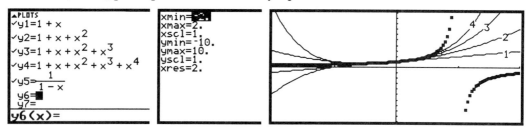

Figure A.3 Graph of four partial sums suggesting convergence in the interval $-1 < x < 1$.

PART V
DIFFERENTIAL EQUATIONS

22. DIFFERENTIAL EQUATIONS AND SLOPE FIELDS

23. EULER'S METHOD

24. SECOND-ORDER DIFFERENTIAL EQUATIONS

25. THE LOGISTIC POPULATION MODEL

26. SYSTEMS OF EQUATIONS AND THE PHASE PLANE

DEMONSTRATION: EULER'S METHOD ON A SYSTEM OF EQUATIONS

22. DIFFERENTIAL EQUATIONS AND SLOPE FIELDS

Often a situation occurs where we know something about a function's rate of change, but the original function may not be explicitly known. For example, we have seen how to antidifferentiate the equation $dy/dt = t$ and solve for y to find $y = t^2 + C$. This is a simple case. It could be that the rate depends on y; the antidifferentiation is not so straightforward then. Equations such as these that involve derivatives are called differential equations.

As in the above example, t is most often the independent variable, as most rates are expressed in terms of time. It is also quite common particularly in textbooks to use the standard x and y variables, with a prime denoting a derivative. For example, the above differential equation could also be written more succinctly as $y' = x$.

A note on the necessity of the TI-92 Plus

In some cases, there are analytic solutions to differential equations, such as the example above. The TI-92 without Plus does not have a more complex differential equation capacity built in. It also does not graph slope and direction fields. If you have reached this far in your calculus without doing so already, it is time to buy the chip and upgrade to a TI-92 Plus.

The solutions of differential equations

In the above example, the analytic solution was written with the constant C. This means that there is a whole set of solutions which differ by a constant: graphing all of these would fill the screen. A graph to show local representative behavior, called slope lines, is called a slope field graph. The main purpose of this chapter is to introduce this tool.

However, we often know specific conditions that determine a single solution, called a particular solution. Particular solutions are graphed as curves. It is common to draw a slope field graph to show the general solution and then superimpose a particular solution curve starting at some initial value.

Discrete vs. continuous representation

The TI-92 Plus is blessed with a differential equation mode in which the continuous differential equation can be directly entered and graphed. It also has a sequence mode that can enter and graph difference equations — a kind of discrete differential equation. The two approaches, the discrete and continuous, offer insight into the topic when considered together. We start with an example of how a discrete difference equation approximates a differential equation.

The discrete learning curve: a sequence function

One theory of learning is that the more you know, the slower you will learn. Let y be the percentage of a task we know. The learning rate is then y' and we assume $y' = 100 - y$. The time unit for this rate is the standard five day work week. The slowing of learning takes place immediately and continuously, but we will start by looking at a discrete model where we assume the same rate all day. This assumption makes the model discrete. In *Calculus*, by Hughes-Hallett, et al., the first example in the differential equations chapter is the following table:

Time (working days)	0	1	2	3	4	5	10	20
Percentage learned	0	20	36	48.8	59.0	67.2	89.3	98.8

Figure 22.1 Approximate percentage of task learned as a function of time.

Consider a new employee who knows 0% of a task at time 0, Monday morning. She learns at the rate $y' = (100 - 0)\%$ during the first day. The part of the task the employee learns on Monday, one fifth of a work week, is

$$y' \cdot (1/5) = 100\%(0.2) = 20\%$$

At this rate, she would have the task entirely mastered in a week. But Tuesday, because she already knows 20% of the task, her learning will slow to

$$y' = 100\% - 20\% = 80\%$$

Therefore, on Tuesday the part of the task she learns is an additional

$$y' \cdot (1/5) = 80\%(0.2) = 16\%$$

In total, she has learned 20% + 16% = 36% by the end of Tuesday.

Each daily total depends on the previous day's total. Writing this as a discrete function, the n-th total will be

$$u(n) = u(n-1) + (1/5)(100 - u(n-1))$$

Making a sequence table

To create a table of this data, we enter **MODE** and change the **Graph** setting to **4:SEQUENCE**. Using ◆_Y= now shows functions called **u1**, **u2**, ..., and below each function is the initial value variable **ui1, ui2**, ...

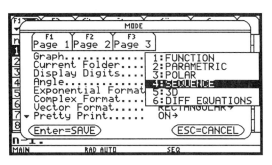

*Figure 22.2 The **Graph** option **SEQUENCE** in the **MODE** menu.*

In Figure 22.3 the function and initial value have been entered. A table is shown starting at 0 with a step of 1. The highlighted table values mean that the employee learns only ≈67% of the task by the end of the week and ≈89% by the end of her second week. According to the model, it is impossible to completely learn the task.

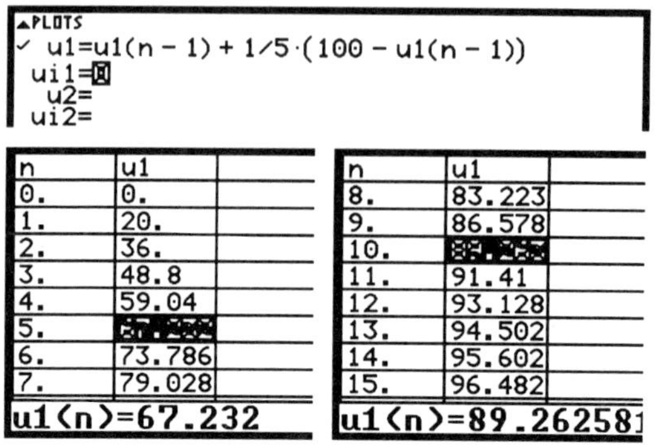

Figure 22.3 Making a table to show values of the discrete learning function and checking the later values by scrolling.

▶ *Tip: With the graphing mode as `SEQ` (shown on the status line), the `TblSet` values define with `tblStart` and `∆tbl` are for n instead of x.*

Making a sequence graph

The graph of a sequence function is produced in the same way as for a "normal" y function. We already have `u1`, so all we lack is setting the window. In Figure 22.4 you see an added set of window variables as we need to set the *n*-values as well as the *x*- and *y*-values. In this case we make the *n*-values match the *x*-values. As with functions, ♦ `_GRAPH` draws the graph which consists of discrete points in `SEQUENCE` mode. The values can be traced with `F3` (`Trace`).

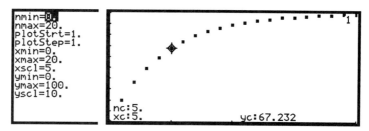

Figure 22.4 Graph (with trace) of the learning sequence function for 20 days.

▶ *Tip: Each graphing mode has its own window settings. Changing `xmin` in `SEQUENCE` mode will not change `xmin` in `FUNCTION` mode.*

The continuous learning curve $y' = 100 - y$

The discrete version of the learning curve assumed that the worker had the same learning rate all day. Now suppose that the learning rate changed every instant,

22. DIFFERENTIAL EQUATIONS AND SLOPE FIELDS   161

i.e., that it is a continuous function. In Figure 22.5, we use MODE to change the Graph type to 6:DIFF EQUATIONS. As with graph types, we enter the equations with the Y= editor. There we see the differential equations are defined using y1', y2', ... and corresponding initial values yi1, yi2, ... In setting the window the t variable for differential equations is like n for the sequence equations. For our graph window settings, we make the t-values the same as the x-values. Recall that this differential equation has weeks as its unit so that xmax=4 will give the same graph as the sequence setting with xmax=20 (since 4 weeks = 20 workdays).

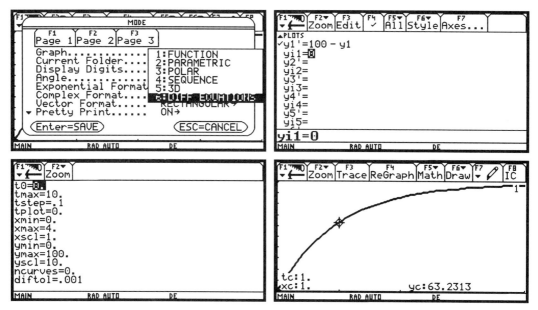

Figure 22.5 *The continuous learning curve and a particular solution with y(0)=0. If your window settings have* **fldres** *at the bottom, your graph will also shows slope lines: we'll discuss this after the next section.*

▶ *Tip: In defining* y1', *use* y1, *not just* y.

▶ *Tip: If your graph shows a set of small lines superimposed on the graph, these are slope lines and for now can be ignored.*

The value of the continuous solution at the end of one week (≈63%) is less than the corresponding discrete value (≈67%). This over-estimate is due to the fact that the discrete model calculates with the same learning level all day long, while the rate is always diminishing in the continuous model.

The analytic solution of a differential equation

You may ask what is the equation of the graph shown in Figure 22.5. That is the essence of solving a differential equation. The TI-92 Plus will give us this kind of symbolic solution in most cases. Unlike the yi' variables in the Y= editor, we enter differential equations on the HOME screen in the traditional mathematical format. The deSolve command can be pasted from the F3 (Calc) menu. When entering a differential equation on the HOME screen, use the ' symbol (the 2nd version of the B key). The initial conditions are added with and, and the independent and dependent variables must be identified. This format is shown in Figure 22.6.

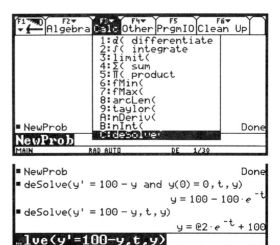

Figure 22.6 The deSolve in the F3 menu. The particular solution of the learning curve differential equation and the general solution.

What if we forget or do not know the initial conditions? This leads to the situation mentioned at the beginning of the chapter where a general solution $y = t^2 + C$ that includes a constant. The TI-92 will identify such constants by @1, @2, etc., similar

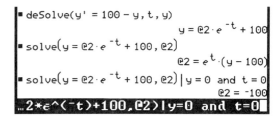

Figure 22.7 The constant of a general solution can be found by using solve and initial conditions.

to what we have seen in some solutions involving periodic functions. These are variables and we can use initial conditions to solve for these values as shown in Figure 22.7. (The symbol @ is the 2nd version of R; remember that pressing ♦_K shows all letter-based 2nd symbols.)

➤ *Tip: You do not have to be in the DIFF EQUATION graph mode to use deSolve.*

Summary of three approaches to differential equations

We have seen three approaches to differential equations. First we used difference equations (SEQUENCE graph mode) to make table values to approximate solutions. Second, in DIFF EQUATION graph mode, we entered the equation and graphed a particular solution. Finally, from the HOME screen, we used deSolve to find a symbolic solution.

Slope fields for differential equations

If we tried to graph all the general solutions, then the entire screen would be covered as our constant changed and the graph shifted vertically. A better approach would be to graph selected particular solutions. A common practice is to refine this further and graph discrete linear pieces that approximate a particular solution. We call such a graph the slope field of the equation. The setting to display the slope field of a differential equation graph is on the GRAPH FORMATS menu, accessed by ♦_F. Figure 22.8 shows selecting the SLPFLD option and the graph of our previous example.

➤ *Tip: The* Fields *option only appears in the* GRAPH FORMATS *menu when the graph mode is* DIFF EQUATION.

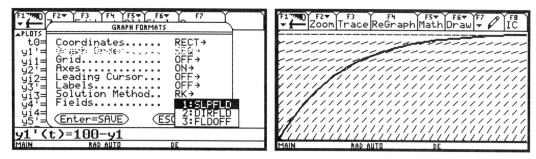

Figure 22.8 Slope field for the learning curve. (The density of your slope field may be different; it is controlled by the WINDOW *setting* fldres *which is* 30 *here.)*

From the slope field, we can trace out particular solutions. For example, it is easy to find the learning curve for an employee with 50% of the skills hired at the beginning of the second week. Pressing F8 (IC) activates a circular cursor that can be moved to the point on the screen where you want a particular solution to begin. In Figure 22.9, you see the circle at (2, 50) showing our choice of initial values.

It may help to summarize the graphing options. In the FUNCTION

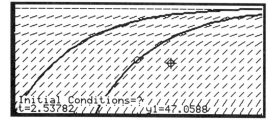

Figure 22.9 Using F8 *(*IC*) to draw additional particular solutions; one has been drawn through (2,50) and the prompt is waiting for another initial value point.*

graph mode, life was simple: just use the 1-2-3 approach of Y=, WINDOW, GRAPH. Now the first thing to do is set the graph mode to one of the six types of graph. As before, define the function(s) in the Y= editor (the function names will alert you to the graph mode). But now, if necessary, include two extra steps before using WINDOW and GRAPH. We have just encountered ♦_F to set graph formats, such as Fields. There is also F7 (Axes…) which makes special axes; this will be used in the Chapter 26.

Slope field for $y' = y$

The solution to this differential equation is the exponential function $y = e^x + C$, although the TI-92 gives this answer in a different but equivalent form in the top, partial screen of Figure 22.10. Graphing shows the slope field and the particular solution that includes $(0,1)$, which is $y = e^x$. This initial point is designated by yi1 and is seen on the screen as a small circle (the same as when we used F8 (IC) above).

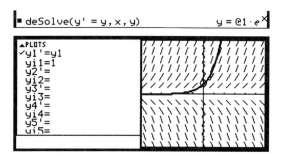

Figure 22.10 *A symbolic solution and the slope field for the equation with a particular solution $y = e^x$. (Using* ZoomDec.*)*

Slope field for $y' = 2x$

The general solution of $dy/dx = 2x$ is $y = x^2 + C$. In the Y= editor, the independent variable x must be translated to t. (This is in contrast to deSolve where any variable name can be used.) By clearing the initial conditions, i.e., leaving yi1 blank, no particular solution is graphed. The first graph in Figure 22.11 is the slope field of this equation. As we can envision from the slope field, a particular solution will be a parabola whose vertex is on the y-axis. Remember that both yi1 and F8 (IC) allow you to graph particular solutions: yi1 from the Y= editor, F8 from the GRAPH screen. In the second graph, we used F8 and chose (0, 0) as the initial value point which resulted in the particular solution $y = x^2$.

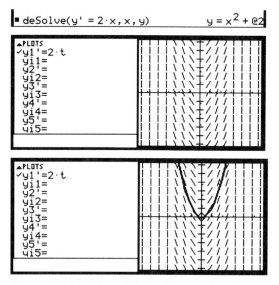

Figure 22.11 *Slope field without and with a particular solution.*

Slope field for $y' = -x/y$ and $y' = x/y$

The deSolve solution in Figure 22.11 says that the equation of a circle, $x^2 + y^2 = $ @3, is a particular solution to the differential equation $dy/dx = -x/y$. (This can also be obtained by implicit differentiation which is shown in Part II Demonstration B.) If you follow a set of slope lines in the first graph, you will trace out a circle centered at the origin. Remember that DE graphing (note the status bar) requires using t as the independent variable and a numbered y as the dependent variable.

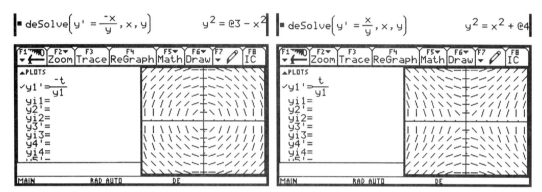

Figure 22.11 Related slope fields.

A sign change from the previous differential equation produces a new slope field in which we could trace out hyperbolas. With a modest leap from the examples in this chapter, you can see that all conic sections can be defined in terms of differential equations.

Slope field for a predator-prey model $y' = (-y + xy) / (x - xy)$

The final example is a slope field of a differential equation that relates two interacting populations (in particular, one population eats the other). Since populations must be non-negative, only the first quadrant of the ZoomDec screen is relevant and you may want to regraph just that region. In that region, a particular solution is an oval curve, as the populations are cyclical, unless we are at (1, 1), a fixed point. Note that the symbolic solution cannot be solved for y. We will look at this differential equation again in Chapter 26.

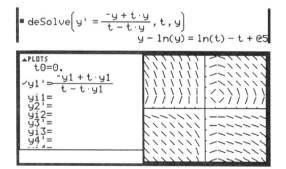

Figure 22.8 Symbolic solution and slope field of the predator-prey model. (Remember to put ✶ between adjacent variables in the definition.)

23. EULER'S METHOD

In the last chapter we saw how the graph of a particular solution follows the slope field. The method of graphing by starting at an initial point and following short segments slope lines is called Euler's method, which we now discuss in more detail. We used this technique to find the discrete values that approximated the particular solution of the learning curve differential equation. To increase accuracy, we can shorten the slope line segments.

The relationship of a differential equation to a difference equation

Suppose we want to find the solution of the differential equation

$$\frac{dy}{dx} = F(x, y) \text{ starting at the point } (x_0, y_0)$$

We think about this continuous differential equation in a discrete way and write

$$\frac{\Delta y}{\Delta x} = F(x, y)$$

or, more explicitly,

$$\frac{y_n - y_{n-1}}{\Delta x} = F(x_{n-1}, y_{n-1}) \text{ with } \Delta x = x_n - x_{n-1}$$

Now solve for y_n and x_n:

$$y_n = y_{n-1} + F(x_{n-1}, y_{n-1}) \Delta x, \quad x_n = x_{n-1} + \Delta x \quad (*)$$

From this, you can see how knowing (x_0, y_0) allows us to find y_1. We will then use y_1 to find y_2, etc.

Translating a differential equation to a difference equation

We now reconsider some of the differential equations from the last chapter as difference equations. First, take the differential equation $dy/dx = y$ starting at (0, 1). With $\Delta x = 0.1$, we make a table using Euler's method and then make a graph to compare it to our analytic solution. This is done in the **SEQUENCE** graph mode. We use **u1** to store the x-values and **u2** to store the y-values. The formulas marked (*) above translate to

(x_n) `u1(n)=u1(n-1)+.1`

(y_n) `u2(n)=u2(n-1)+u2(n-1)(.1)=1.1*u2(n-1)`

We want to compare values on the interval $0 \leq x \leq 1$, so with $\Delta x = 0.1$ we will need to calculate 10 values, shown in Figure 23.1.

23. EULER'S METHOD   167

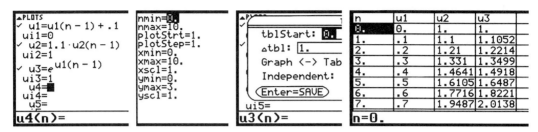

Figure 23.1 Making a table to compare the approximation by Euler's method to the particular solution of the differential equation y′ = y.

We set u3=e^(n/10) to be the exact solution we found in the last chapter, $y = e^x$. (Note $x = n/10$.) We compare the numeric approximations in u2 to the true values in u3 by using TABLE. The Euler method approximation for an x-value such as .3 is pretty good, but you can see the error increase as x increases. If needed, these approximations can be improved by making Δx smaller. In Figure 23.2 we graph u2 and u3 together for comparison. The x-scale is set $0 \leq x \leq 10$ to match the n-scale for graphing, but in fact $x = n/10$.

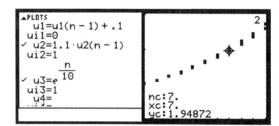

► *Tip: It is possible to use u1 as the horizontal axis and have the axis be true scale; this is set with F7 (Axes...) which we will use in Chapter 26.*

Figure 23.2 Making a graph to compare the approximation by Euler's method to the particular solution of the differential equation y′ = y. Use Trace to identify values.

Euler's method for y′ = -x / y starting at (0,1)

The TI-92 knows methods for solving differential equations numerically. The default setting is RK which stands for the Runge-Kutta method. This is slower than the Euler method, but it is generally more accurate (see any differential equations text for details). There is also EULER.

Consider the differential equation $dy/dx = -x/y$ starting at (0,1). We want to investigate the effect of using a smaller value of Δx by graphing the solution on the interval $0 \leq x \leq 1$ first with $\Delta x = 0.1$, then with $\Delta x = 0.01$. This time we use the built-in Euler setting. The setup process in Figure 23.3 follows the steps outlined in the last chapter.

In MODE, change the Graph setting to 6:DIFF EQUATIONS.

Press Y= and enter the equation. Remember to use t and y1. The initial condition is set with yi1=1.

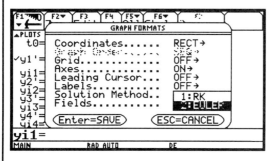

Press ♦_F to access the GRAPH FORMATS and set the Solution Method to 2:Euler. Also, set Fields to 3:FLDOFF (this step is not shown).

We are interested in $0 \leq t \leq 1$. With a variable change from x to t, now $\Delta x =$ tstep=.1. The x- and y-values are selected to show near equal scale. The iterations between each tstep is given by Estep; 1 is its default.

Use TRACE to find the points (0, 1), (0.1, 1), (0.2, 0.99), (0.3, 0.97), etc.

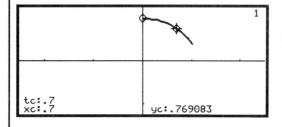

23. EULER'S METHOD   169

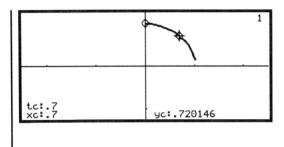

Now increase `Estep` to 10 to improve the accuracy of the approximation: traced points include (0, 1), (0.1, 0.995), (0.2, 0.981), (0.3, 0.956), etc. The graph is more circular, as it should be (recall the analysis of this differential equation in the previous chapter).

Figure 23.3 Using the built-in `Euler` method.

▶ *Tip: Using `tstep=.01` with `Estep=1` produces the same numeric values as `tstep=.1` with `Estep=10.`, so we did in effect graph the solution with $\Delta x = 0.01$. Changing `Estep` instead of `tstep` makes graphing and tracing faster.*

Euler gets lost going around a corner

Use Euler's method with caution. As you wander away from your initial point, you may encounter increasing errors. The next example shows that there are some paths where we encounter an infinite slope that will cause problems.

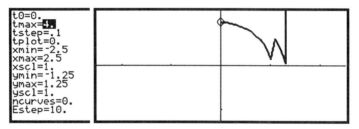

Figure 23.4 Euler's method breaks down as at (1,0) where the slope is undefined.

Let's take the previous example and increase `tmax` in hopes of finishing out the first quadrant. Figure 23.4 shows the folly as we come to (1,0). Changing the setting from `Euler` back to the default `RK` eliminates this hazard as shown in Figure 23.5. Unless investigating Euler's method, use the `RK` setting.

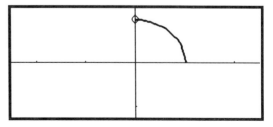

Figure 23.5 Graph of the same settings after changing `Solution Method` to `RK`.

24. SECOND-ORDER DIFFERENTIAL EQUATIONS

A second-order differential equation is one that has a second derivative in its expression. Before making the transition from a symbolic equation to calculator definitions, we rewrite these equations by solving for the second derivative. Thus we will find it easiest to consider a differential equation in the form

$$y'' = F(x, y, y')$$

As before, the TI-92 will give symbolic solutions of these differential equations, but it can also be used to get numerical solution values and to check your analytical solution graphically. We begin with the simplest second order equation.

The second-order equation $s'' = -g$

A classic equation from physics is,

$$\frac{d^2s}{dt^2} = -g$$

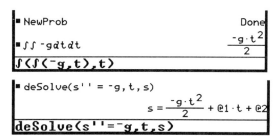

Figure 24.1 *The TI-92 does nested integration and solves second-order equations.*

where g is the constant force of gravity on a falling object, s is displacement in feet, and t is time in seconds. If we assume the initial velocity is zero ($v_0 = 0$) and the initial distance is zero ($s_0 = 0$), anti-differentiating once gives $s' = -g \cdot t$, and then again gives the solution $s = -g \cdot t^2/2$. The first calculation in Figure 24.1 confirms our answer and shows that the TI-92 allows nested integration. The second calculation uses the `deSolve` which gives the general answer with constants `@1` and `@2`.

Now let's investigate the `DE` graph mode to produce a graphical solution. We will also make this example more down to earth by assuming that the gravity constant g is 32 feet/sec^2.

There is no `y1''=` menu item for entering second-order equations in the `Y=` editor, but we can build a second order equation as a system of two first order equations:

$$\text{y1'=y2 and y2'= -32 give}$$

$$\text{y1'' = (y1')' = (y2)' = y2' = -32}$$

➤ *Tip: The notation for* `deSolve` *is more straightforward than* `Y=` *definitions.*

24. SECOND-ORDER DIFFERENTIAL EQUATIONS

For our first example, we review in detail the sequential process we use to graph solutions of differential equations. Following this prescribed sequence, you can obtain the other examples in the chapter. The five steps to set up a differential equation graph are

- Set **MODE** to **DIFF EQUATIONS**
- Define the equation with **Y=** editor (include initial conditions)
- Set **GRAPH FORMAT** (**♦_F**) **Fields** to **FLDOFF**
- Set **WINDOW**
- Draw **GRAPH**

These steps have been followed to produce the graph in Figure 24.2. After defining the second-order equation as a set of two first-order equations, deselect all but the first function to be graphed.

Since we know this solution is $y = -16x^2$, we set t to act like x and include negative y-values. The **diftol** setting indicates the **RK** setting (this value prescribes the error tolerance in the Runge-Kutta algorithm), otherwise **EStep** (Euler step) would be shown.

Once the graph is drawn, you can use **DRAW**, **TRACE** or **F8** (**IC**) to investigate further.

➤ *Tip: The one-step* **Zoom** *options (***ZoomStd***, ***ZoomDec***, etc.) are rarely useful for differential equations since they are centered at the origin.*

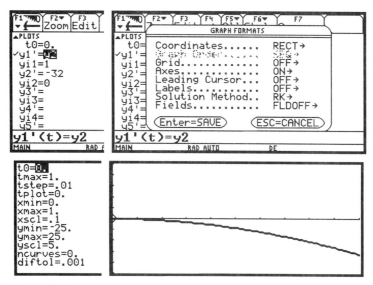

Figure 24.2 Writing a second-order equation as a set of first-order equations.

The second-order equation $s'' + \omega^2 s = 0$

Next we look at a second-order differential equation that depends only on the dependent variable.

$$\frac{d^2s}{dt^2} = -\omega^2 s, \text{ where } \omega > 0$$

This differential equation describes simple harmonic motion. We see in Figure 24.3 that its solution is

$$s(t) = C_1 \cos \omega t + C_2 \sin \omega t$$

As a specific example, we will $\omega = 2$ and set the initial conditions to be $s(0) = 1$ and $s'(0) = -6$. In `deSolve`, the initial conditions modify the general equation by using `and`. This is shown in Figure 24.4 where we see that the particular solution has constants $C_1 = 1$ and $C_2 = -3$.

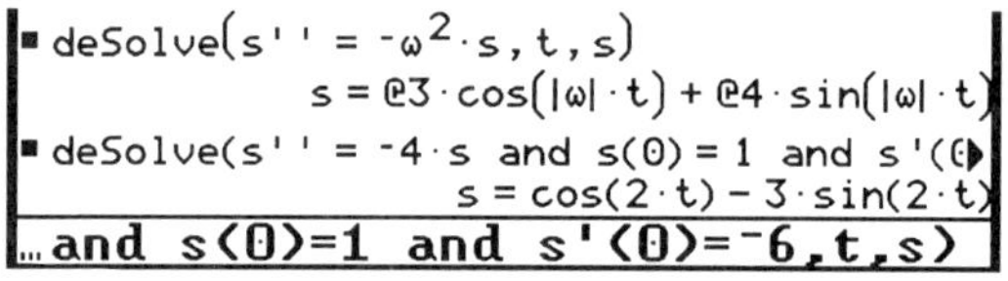

You may wonder why we would want to bother using the `DE` graph mode to find a graphical solution when we already know the particular solution equation and could graph it in the `FUNCTION` mode. The answer is that, by graphing in `DE` mode, we can add the slope field, enter new initial conditions, and compare graphs. Let's graph an approximation and compare it to this known solution. Since this is a trigonometric solution, we graph with $0 \le x \le 2\pi$. The default style of a `DE` equation is `Thick`, but for comparing to the analytic solution we change to `Dot`. The `DE` graph is shown in Figure 24.5.

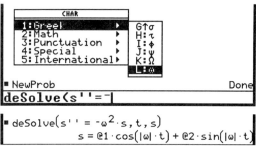

Figure 24.3 General solution of a second order equation with a Greek letter constant ω from `2nd_CHAR` menu (also available through `2nd_G_W`).

Figure 24.4 A particular solution for simple harmonic motion. Initial conditions are entered with `and`.

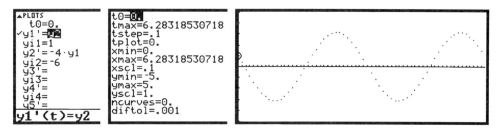

Figure 24.5 The graph (`Dot` style) of the solution to $s'' = -4s$ with initial conditions $s(0) = 0$ and $s'(0) = -6$. Notice variable name conversion from s to `y1`.

24. SECOND-ORDER DIFFERENTIAL EQUATIONS

By using **F6** (**Draw**), the analytic solution is added to the graph in Figure 24.6. Notice that this formula must be entered with x as the independent variable. We see no noticeable difference between the graphs.

The linear second-order equation $y'' + by' + cy = 0$

Equations of this type are called linear second-order equations because isolating the second derivative gives a linear equation with variables y and y' and coefficients b and c. One application of this general equation is describing the motion of a spring.

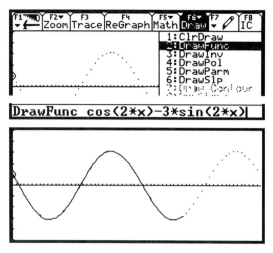

Figure 24.6 Use DrawFunc to graph a solution and compare it to the original graph.

The solution of this second-order linear equation hinges on the sign of the $b^2 - 4c$ inside the square root in the exponent. The solution shown in Figure 24.7 can be written as

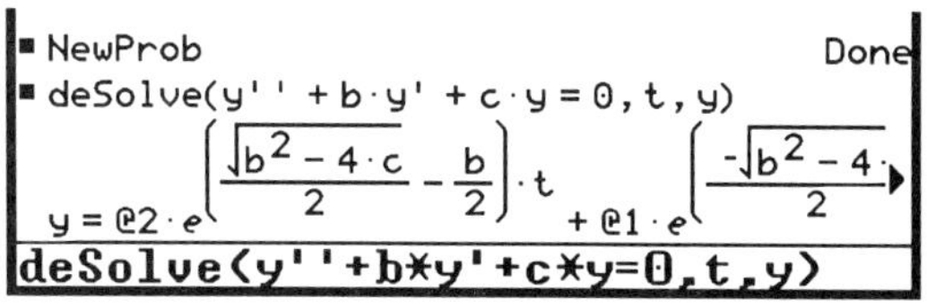

Figure 24.7 The general solution of the linear second-order equation.

$$y = C_1 e^{r_1} + C_2 e^{r_2} \text{ with } r_1 = \frac{\sqrt{b^2 - 4c}}{2} - \frac{b}{2} \text{ and } r_2 = \frac{-\sqrt{b^2 - 4c}}{2} - \frac{b}{2}$$

These two values are the zeros of the equation $r^2 + br + c = 0$. This quadratic is often called the characteristic equation for the differential equation. We consider the three possible cases: the roots are real and distinct, the roots are repeated, and the roots are imaginary.

The overdamped case: $y'' + by' + cy = 0$ with $b^2 - 4c > 0$

Consider the second-order linear equation with $b = 3$ and $c = 2$, so that the discriminant is positive, i.e., $3^2 - 4(2) > 0$. The general solution and a particular solution with initial conditions $y = -.5$ and $y' = 3$ at $t = 0$ are shown in Figure 24.8. Thinking of

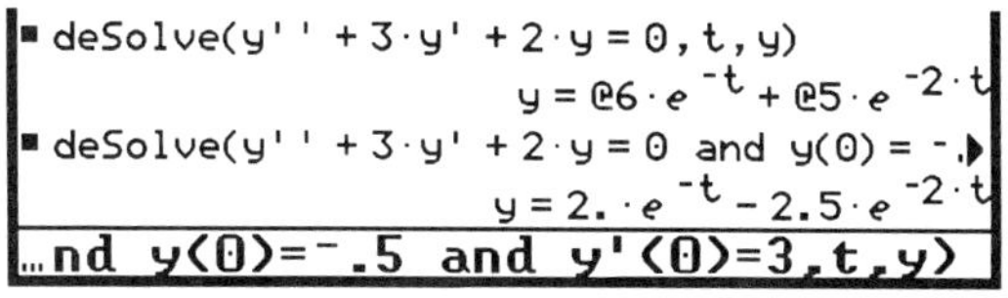

Figure 24.8 The characteristic equation has distinct real roots.

this as a spring's motion in oil, we see that it starts half a unit below equilibrium, swings past equilibrium ($y = 0$) and then its motion is damped until the spring is at rest.

The power of DE graphing is that we can explore new initial conditions. In Figure 24.9 we see the setup that produces the graphical solution (be careful to select only y1'). In the bottom screens we press F8 (IC) and are met with a new dialog box; just press ENTER and a circular cursor will allow you to set new initial conditions. By moving the cursor to (0,0) and pressing ENTER, a second graph is drawn. In this case, the spring starts at equilibrium and appears to never swing back past the equilibrium as it comes to rest.

How can we be sure that the spring will not oscillate through equilibrium as it

Figure 24.9 DE graph of the solution and new initial conditions to create a second solution.

comes to rest? The Zero option cannot be chosen from the DE graph mode. So in Figure 24.10 we symbolically find the zeros of the analytic solution from Figure 24.7 (be certain to remove y= and add ,t) when you paste the equation into the zeros command). The when in the answer gives a condition for the one zero to occur; recall from Figure 24.8 that the constants for $b = 3$ and $c = 2$ are 2 and -2.5, which satisfy the condition.

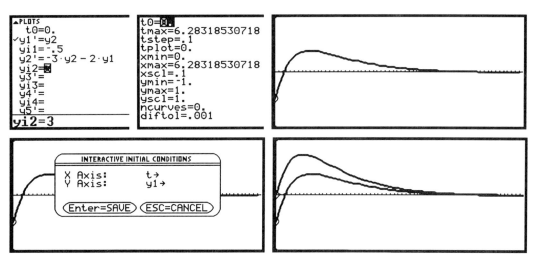

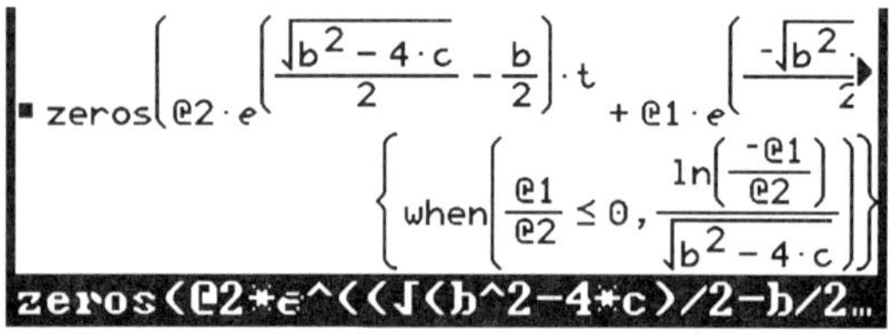

Figure 24.10 Symbolic verification of our graphical hypothesis that the spring swings past equilibrium only once.

The critically damped case: y″ + by′ + cy = 0 with $b^2 - 4c = 0$

To find the general solution when the discriminant in zero, we return to our original equation and add the restriction that $c = (b/2)^2$. This gives the solution in Figure 24.11. For an actual example, suppose $b = 2$ and $c = 1$ (so $2^2 - 4(1) = 0$). Figure 24.12 shows the graph of the particular solution with the same initial conditions, $y = -.5$ and $y' = 3$ at $t = 0$, and window as in the last example. The graph is similar to the solution of the previous case, but notice that it is not damped as quickly.

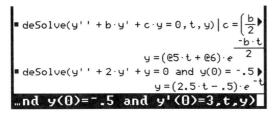

Figure 24.11 The critically dampened case with c in terms of b and a particular solution with b = 2 and c = 1.

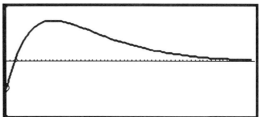

Figure 24.12 The critically dampened graph (similar to the overdamped case of Figure 24.9).

➤ *Tip: When changing y2′ for reproducing Figure 24.12, don't forget to deselect it.*

The underdamped case: y″ + by′ + cy = 0 with $b^2 - 4c < 0$

The TI-92 will not usually simplify expression when inequalities are used as modifiers, so we get no new solution form when we use $b^2 - 4c < 0$ as a conditional statement. The general symbolic solution in this case is known to be

$$y(t) = C_1 e^{\alpha t} \cos \beta t + C_2 e^{\alpha t} \sin \beta t$$

where $r = \alpha \pm i\beta$ are complex zeros of the characteristic equation.

When the parameters b and c given actual values, **deSolve** will produce the solutions. For example, suppose $b = 2$ and $c = 2$, then the discriminant $2^2 - 4(2) < 0$. Figure 24.13 shows that in this case $\alpha = -1$ and $\beta = 1$ and the **deSolve** answer fits the symbolic solution above. We then add initial values to show the particular solution when the spring is 2 units above equilibrium and released with no velocity (i.e., $y(0) = 2$ and $y'(0) = 0$).

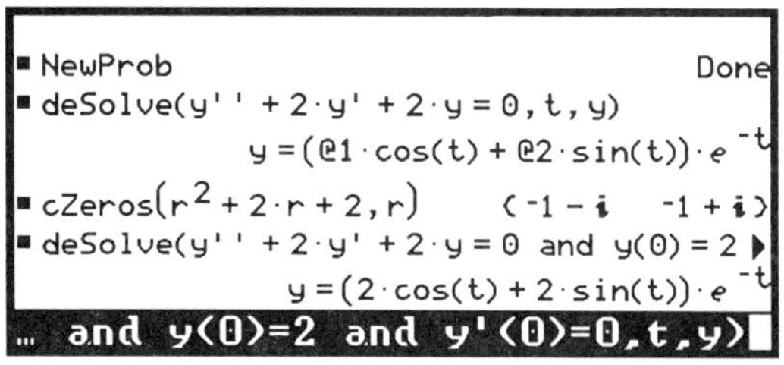

Figure 24.13 General and particular solutions which use the real and complex parts of the roots of the characteristic equation.

For comparison to the other two damped cases, in Figure 24.14 we use DE graph mode to show a particular solution with the previous boundary conditions in the same window. We see that the motion is damped but crosses the equilibrium point at least twice in that interval. From the solution equation, we know that this graph is an exponentially damped sine curve and so it will continue to periodically cross the equilibrium line ($y = 0$); the symbolic solution in Figure 24.15 shows the closed form of the repeating zeros.

Figure 24.14 Graph of the particular solution with initial conditions $y = -.5$ and $y' = 3$ at $t = 0$.

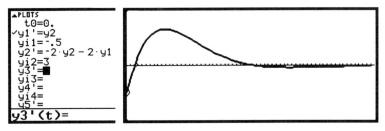

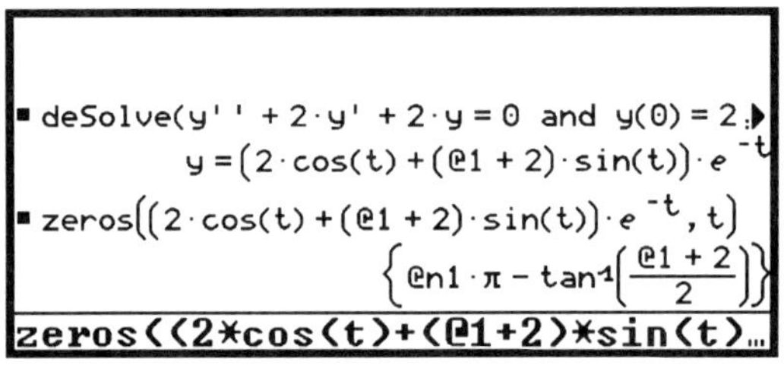

Figure 24.15 Symbolic solution to when this spring crosses equilibrium.

25. THE LOGISTIC POPULATION MODEL

One of the remarkable features of a TI-92 is its statistical capability. In this chapter we introduce the **Data Editor** application and show data graphed as a scatter plot. We will develop a model for the population growth of the United States with a logistic regression curve. The calculator's statistical features can be more fully explored in Chapter 9 of the TI-92 *Guidebook*.

Then we will derive the logistic equation by using a differential equation. For this second part it is suggested that the reader refer to the section "Models of Population Growth" in the differential equations chapter of *Calculus* by Hughes-Hallett, et al. This material will parallel that exposition and show how the calculations and graphs can be shown using a TI-92 Plus.

Using the **Data Editor** for entering US population data

The first step is to enter the data from the US Census Bureau. Typically, annual data are not indexed by the year itself, but by years from a base year. Also, large numbers are usually rounded. Our base year is 1790 and we will round populations to the nearest tenth of a million. For example, the data pair for 1800 is (10, 5.3). The data are from 1790 to 1940; later we see how well they predict more recent growth. Because this is the our first presentation of data lists and statistical regression, we return to a more explicit style with a long, annotated sequence of screens.

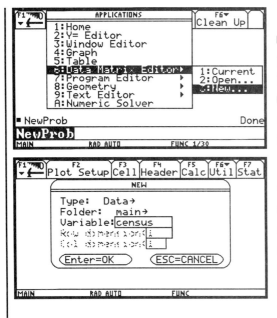

To create a new data file, press APPS 6 3:New... The **Data Editor** presents a dialog box. When returning to the last used data file, you can press APPS 6 ENTER. If you are looking for a specific data file, use APPS 6 2:Open to find it.

In the dialog box we give the file a variable name. Dialog boxes require up and down arrows to move among items and the right arrow to produce item choices. You must press ENTER after making an entry. Then you must press ENTER again before you exit to OK all the current dialog box settings.

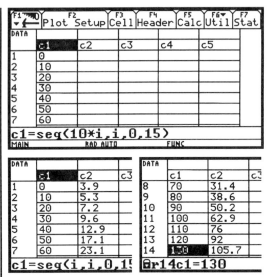

The years can be entered sequentially, but it is easier to press F4 (Header) and enter the seq command shown. The columns act as lists named c1, c2, etc. The row numbers are shown at the far left. Enter the US population data that are shown below.

US population data from 1790 to 1940, rounded to tenths of a million.

Figure 25.1 Creating the census *data structure and entering the data.*

Using Plots to graph data lists

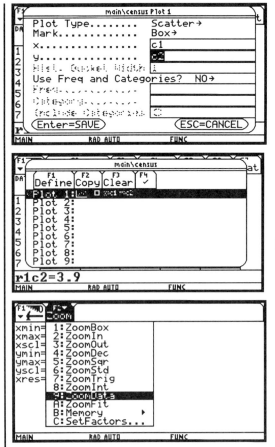

To look at the data graphically, we press F2 (Plot Setup) and then F1 (Define) to reach the dialog box shown. Arrow down to x and type c1 ENTER then arrow to y and type c2 ENTER. Then press ENTER again to save these settings.

This dialog box holds a summary of plot assignments for graphing data lists. Here plots are selected/deselected and, for each plot, it is the entryway (using F1) to a screen like the previous one in which actual definitions are made.

We need to set a graph window. Press ♦_WINDOW and F2 (Zoom). A handy zoom setting is 9:ZoomData which sets a window that fits all your data.

25. THE LOGISTIC POPULATION MODEL   179

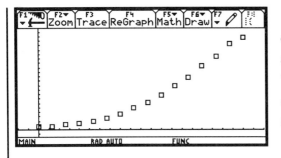

The `ZoomData` selection automatically graphs the data points. You can also use ♦_GRAPH at anytime to graph. You will have the `yi` function graphs on your screen if you did not turn them off.

Figure 25.2 Creating a scatter plot of the data.

Fitting data with a logistic equation

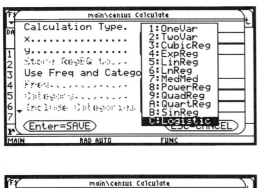

We return to our original data screen with APPS 6 1 (shown in background). Press F5 (`Calc`) and arrow right to see the `Calculation Type` menu. The choices from 3 to C fit data lists with an equation (the TI-92 without Plus stops at A). From the look of our data, we choose C:`Logistic`.

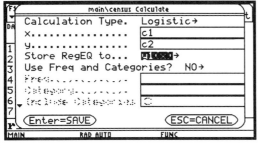

The logistic equation is a functional relationship between two variables. We identify `c1` as `x` and `c2` as `y`. Just below this entry is an option to store the equation as a function. We select `y1` to store the derived equation for later graphing.

After pressing ENTER, a `busy` indicator will show until the calculations are completed. The coefficient results we obtain here will be very close to those we will now derive by using a differential equation.

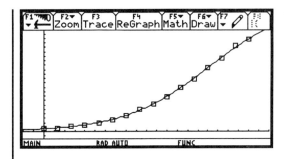

Now press ENTER to clear the STAT screen and ♦_GRAPH to see how well the logistic equation fits the given data. If you look in the Y= editor, you will see that both Plot 1 and y1 are selected.

Figure 25.3 Finding and graphing the logistic regression curve.

The logistic curve

Logistic data can be spotted from a graph because it starts like a concave up exponential function and then takes on concave down exponential behavior. There is a more formal approach: its relative growth rate is linear. We investigate our model for this behavior.

Estimating the relative growth rates: *P' / P*

We want to find the relative annual growth rate, but we only have data for decades so we settle for an approximation that uses data from the previous and next decade:

$$\frac{1}{P} \cdot \frac{dP}{dt} \approx \frac{1}{P_i} \cdot \frac{P_{i+1} - P_{i-1}}{20}$$

Figure 25.4 shows the calculations from the following seq command

```
c3=seq((c2[i+1]-c2[i-1])/20/c2[i],i,2,15)
```

that will do all these calculations and store them in c3. Beware that we had to start at I=2 and end at I=15 because we needed a previous and next decade for each of our calculations. Looking at the first six c3 values, we see that a rough estimate for the relative growth rate is 3%.

DATA	c1	c2	c3	c4	c5
1	0	3.9	.03113		
2	10	5.3	.02986		
3	20	7.2	.02969		
4	30	9.6	.02907		
5	40	12.9	.02982		
6	50	17.1	.03095		
7	60	23.1	.02468		

c3=seq((c2[i+1]-c2[i-1])/20/(...

Figure 25.4 Using the seq command to estimate the relative growth rate and storing it in c3.

25. THE LOGISTIC POPULATION MODEL

Modeling population growth with a simple exponential function

A naîve model builder would say the population starts at 3.9 (million) and grows at a 3% rate so that the following simple exponential model could be used: $y = 3.9e^{0.03x}$.

But if we scroll down the table we see that the relative growth rate declines. By looking at the estimated relative growth rates for the years from 1860 to 1930, we see that the growth rate slows. We now want like to quantify this decline by comparing $(1/P)(dP/dt)$ to P. This requires some data rearrangement. The estimated relative growth rate is in c3 but, because of the way it was calculated, it has no entry for the first or last decade (it has only 14 terms while there are 16 population entries). We make a new list, c4, with the populations corresponding to the c3 list by loading a copy of c2 into c4 and then deleting the first and last entries. (It would not do to delete entries from c2 since the calculations for c3 use all of the c2 data.) To copy a list, move the cursor to c4 and press F4 (Header), then type c2 (the name of the list to be copied), and press ENTER. Because we want to alter the list, we must clear the c4 definition with backspace (←), otherwise it cannot be edited since it is defined to be an exact replica of the current c2. Deleting this formula will leave the current entries unaltered. In Figure 25.5 we see the copy of c2 in c4 and then the result of using backspace (←) to delete the first and last entries of c4. The relative growth rates in c3 now correspond to the populations in c4.

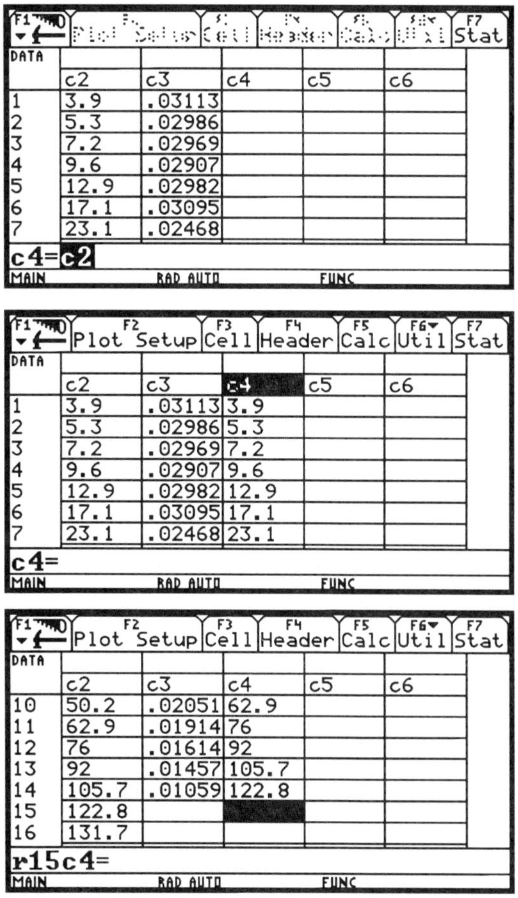

Figure 25.5 Copy c2 to c4 and delete the first and last entries of c4 so it corresponds to c3.

A scatterplot of P' / P against P

We now create a scatterplot to look for a relationship between the relative growth rate data (c3) and the corresponding population data (c4).

In Figure 25.6, `Plot 2` has been selected, turned on, and selected as a scatterplot for `c4` and `c3`. We use `ZoomDec` to reset the window from the previous data fit.

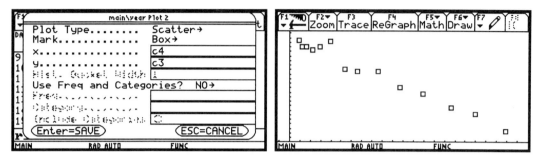

Figure 25.6 Scatterplot of `c3` against `c4`.

▶ *Tip: A frequent annoyance is that a* `Y=` *function is left on from previous use and shows up on the graph screen. Even if it is outside the window, it still slows the graphing. Check the* `Y=` *screen for proper selections before graphing. Using* `F6` *(`Clean Up`) `2:NewProb` will turn off all functions and plots.*

▶ *Tip: The status of a plot as on or off is shown at the top of the* `Y=` *edit screen: plots that are on are checked. As with function, the status can be toggled with* `F4`.

Finding a linear regression line to fit the data

Our scatterplot in Figure 25.6 has a linear look, so we fit it using a linear equation. We have already seen the TI-92 find a logistic equation based on data lists. Regression equation is the statistics term, but for our purposes we will think of it as meaning a *best fit* type of equation. When we say regression line equation, we mean the equation of the line that comes closest to fitting our data, just as our logistic regression equation fit the population data nicely.

Figure 25.7 shows the use of with `c4` and `c3` specified to find the equation of the line $y = -.00017x + .0318$. By putting this equation in `y2` and graphing (with `Plot2` on), we see that the regression line fits this relationship fairly well (how well is measured by the correlation coefficient `corr` which is very near -1).

Using the regression line to rewrite the differential equation

Now we transform the linear regression equation $y = ax + b$ back to its differential equation form with $y = (1/P)(dP/dt)$ and $x = P$. Next we multiply by P to get the differential equation

$$\frac{dP}{dt} = 0.0318P - 0.00017P^2$$

25. THE LOGISTIC POPULATION MODEL

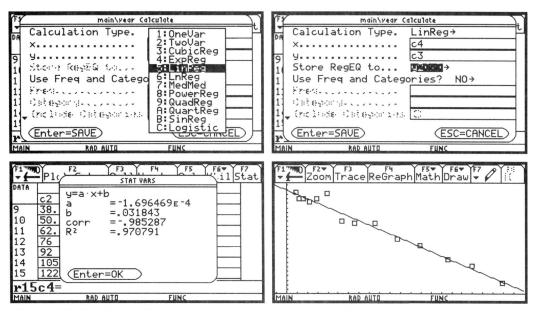

Figure 25.7 Finding and graphing the regression line to the plot.

And then we use `deSolve` as shown in Figure 25.8. For completeness, the end of the `deSolve` command (with its initial value `P(0)=3.9`) is shown on the entry line. The answer is given as `1/P`, so we use `solve(...,P)` to get an explicit function in terms of `P`. You can divide both numerator and denominator by the exponential term and convert it to exponential notation. A rounded equation in traditional form is

$$P = \frac{187}{1 + 47e^{-0.0318t}}$$

This is quite close to the logistic regression equation given in Figure 25.3, which with rounding is

$$P = \frac{181.4}{1 + 51.1e^{-0.0329t}} + 0.6$$

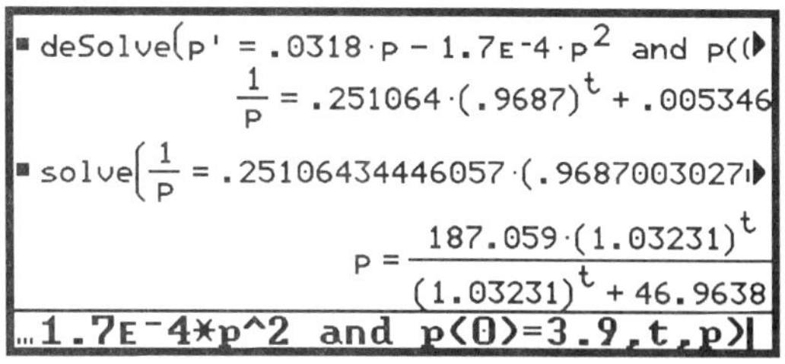

Figure 25.8 The solution from `deSolve` solved for P.

You can graph either of these equations or use DE graph mode with

$$\text{y1'}=.0318\text{y1}-.00017\text{y1}^{\wedge}2 \text{ and } \text{yi1}=3.9$$

to find whether we have a good predictor for our current growth. The predictions are lower than the actual numbers, principally because of the baby boom.

▶ *Tip: Using the proper regression equation to model a situation is important. The various STAT CALC regression options allow you to try out different models on the same data to look for a good equation.*

26. SYSTEMS OF EQUATIONS AND THE PHASE PLANE

In this chapter we look at examples of systems of differential equations where the independent variable is time. Any two of these equations can be graphed to show their relationship to each other. Points on the graph are called the phase trajectory or orbit in the phase plane. Two popular examples of using systems of differential equations are the S-I-R model the predator-prey model.

For example, in a predator-prey model we can graph the predator population with time on the x-axis and then also graph the prey population with time on the x-axis. We can also create a phase plane graph with predator population on the y-axis and prey population on the x-axis.

As in the previous chapter, our goal here is to use the calculator's graphics to supplement detailed examples from *Calculus*, Hughes-Hallett, et al. Refer to that text for further background on the derivation and interpretation of the models.

The S-I-R model

The S-I-R stands for Susceptible - Infected - Recovered, so you can tell that this is used to model epidemics. The population is divided into the three groups and people move from S to I to R, or they just stay in S. The three rates in terms of time are

$$\frac{dR}{dt} = bI$$ the recovery rate depends upon the number infected

$$\frac{dS}{dt} = -aSI$$ the susceptible rate is negative and depends on both the number of infected and susceptible

$$\frac{dI}{dt} = aSI - bI$$ the infected rate is the negative of the sum of the other two rates.

Because knowing any two of the quantities S, I, R will automatically determine the third, we will concentrate on the last two rates.

The boarding school epidemic

The following example can be found in more detail in *Calculus*, Hughes-Hallett, et al., page 563. There were 762 students in a boarding school and one returned from vacation with the flu. Two more students became sick the second day. This means we can approximate a = -0.0026 from

$$\frac{dS}{dt} = -aSI \text{, since } (-2) = a(762)(1)$$

26. SYSTEMS OF EQUATIONS AND THE PHASE PLANE   185

This flu lasts for a day or two, so we will assume half of the sick get well each day (thus $b = -0.5$).

Time plots for the model

In Chapter 24 we listed five steps to set up a differential equation graph. In this chapter we sometimes need a another step to change the assignment of the axes. After defining the equations, use F7 (Axes) to check and, if necessary, change the x- and y-axes. In all previous examples we have used the default setting TIME (the axes are t and y), but after this first example we will need to change the setting from TIME to CUSTOM. Because of these impending changes, we will set LABELS to ON with ♦_F. Our sequence is now:

- Set MODE to DIFF EQUATIONS
- Define the equation with Y= editor (include initial conditions)
- Set AXES with F7 (Axes)
- Set GRAPH FORMAT with ♦_F
- Set WINDOW
- Draw GRAPH

In Figure 26.1 we define the differential equations and graph the variables S (y1) and I (y2) over time. With the axes set to TIME, both y1 and y2 will be graphed against time on the horizontal axis. We want two graphs (and no slope field), so we use ♦_F to select FLDOFF and we turn the axes Labels ON. For the

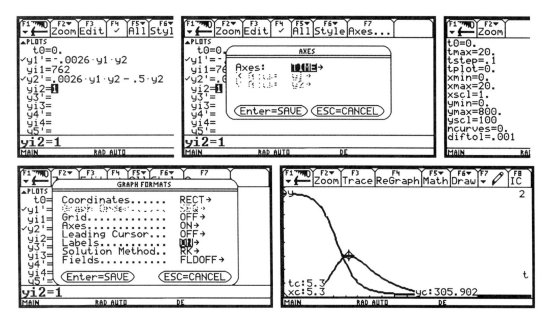

Figure 26.1 The graphs of S and I over time. We see that the susceptible are in a constant decline and the infected peak in the sixth day.

window, we look at the epidemic for 20 days on the horizontal axis and, as there are just under 800 students, we set ymax=800. Tracing gives an estimate at the maximum number of infected students.

Phase plots for the model

In Figure 26.2 we look at the *SI* phase plane. The same definitions of y1' and y2' and initial conditions remain from the previous plot, but now we change the AXES to CUSTOM and use the default settings, X Axis:y1 and Y Axis:y2. The model requires $0 \leq y1 \leq 800$, $0 \leq y2 \leq 400$ and $0 \leq t \leq 20$ days. Watching the screen as it graphs, you will notice that the graph is drawn from right to left, as the first point is $(S, I) = (762, 1)$. Using TRACE on this phase plot feels backward because the right arrow key will increase *t* which causes the trace cursor to move left. If you press the left arrow key at the start, then nothing will happen. You can use TRACE to find the peak at approximately (192, 306), where $t = 5.3$ days. The maximum value is of interest in the next section.

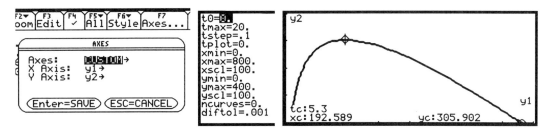

Figure 26.2 The SI phase plane. Use F7 (Axes) in the Y= editor for CUSTOM settings.

➤ *Tip: The TRACE cursor moves through the points connected to the t-values when the right arrow is pressed; the cursor movement is not necessarily from left to right. Moving backward through time is much slower than moving forward.*

➤ *Tip: Pressing the 2nd key before the right arrow puts it in turbo mode, making the cursor jump further.*

Direction fields for the S-I-R model

Like the slope fields for functions in terms of *t*, we can graph slope lines for the relationship of *I* to *S*. Unlike a slope field which has *t* on the *x*-axis, a direction field shows a relationship between two quantities, like y1 and y2, in a system. Time is not shown on either axis, but each point does have a time value implicitly attached. This can be seen when you trace a phase plot; the cursor starts at tmin and increases by tstep.

We use ♦_F to set DIRFLD on the bottom row. The resulting graph is shown in Figure 26.3. Of particular interest is the fact that along any particular solution, the peak (threshold value) is at the same value (S = 193) on the horizontal S-axis. But if S is less than 193, then the I-values decrease immediately. We use F8 (IC) to explore another phase plot at a smaller school (y1 = 500) with a one student (y2 = 1) beginning the outbreak.

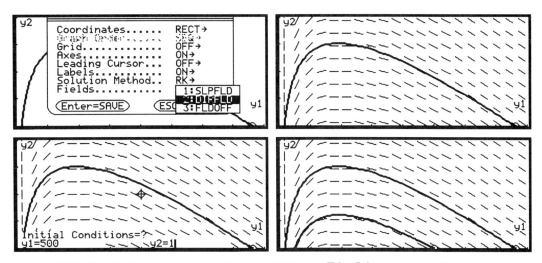

Figure 26.3 *The SI phase plane with direction field. Use* **F8** *(**IC**) to graphically explore particular solutions with other initial conditions.*

➤ *Tip: Pressing* **F8** *(**IC**) will activate a circular cursor to move to the initial conditions, but it is much easier and more accurate to type in the numeric values rather than use the arrow keys.*

Predator-prey model

In this kind of model, we let x be the number of predators and y the number of prey. The details for this kind of system are found on page 244 of the TI-92 *Guidebook* where analysis is done using the sequence graphing. The TI-92 Plus is much more powerful and can deal with this model using continuous functions. In our example, we simplify greatly, setting all constants to 1. This follows the approach taken in *Calculus*, Hughes-Hallett, et al., page 567.

The simplified predator-prey system is

$$\frac{dy}{dt} = y - yx \text{ and } \frac{dx}{dt} = -x + yx$$

We will define the predator population *x* as y2 and the prey population *y* as y1 (with units in millions). We follow the presentation order of the previous model, first graphing the two populations over time.

Time plots for the predator-prey model

In Figure 26.4 the individual predator and prey differential equations with respect to time are entered and the plot style of the first equation is set to Line. (The default style for differential equation graphs is Thick.) Although initially you may have no idea how to set the window, a little trial-and-error can lead to the window shown in Figure 26.4. We made it wide enough to show the periodic nature of the two graphs.

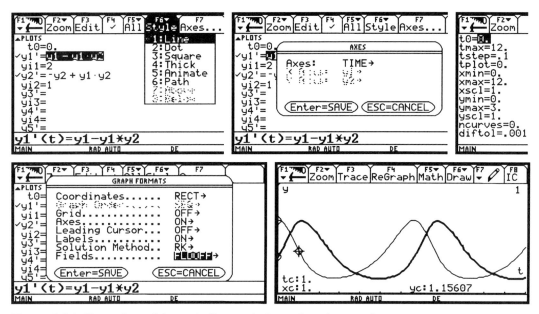

Figure 26.4 Time plots of the periodic populations of predator and prey.

Phase plots for the predator-prey model

We now change the AXES and set the window to match the size of the predator and prey numbers. From the previous graph, we know that more than a complete cycle will take place if tmax=12, so we leave that setting. (We could estimate a smaller value of eight by looking from peak to peak on the graph.) Set xmax=3 since both populations are less than 3 (million). Remember when tracing values to press the right arrow key even though the trace cursor will move left at first; you are moving up in *t* which may correspond to moving right or left in *x*.

➤ *Tip: When graphing a phase plot, the curve will have the style specified by the function defined on the x-axis.*

Direction field for the predator-prey model

By changing the ♦_F Fields setting to DIRFLD, we can superimpose the direction field over the phase plot. What we see from this direction field in Figure 26.6 is an equilibrium point at (1, 1). At these values, the populations will theoretically remain stable and not have the kinds of cycles that we saw in Figures 26.4 and 26.5.

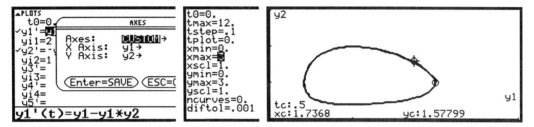

Figure 26.5 Graphing prey against predator.

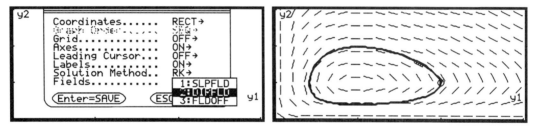

Figure 26.6 The direction field over the phase plot of Figure 25.5.

The direction field of this system of two differential equations might look familiar; it was part of the last example in Chapter 22, but as the slope field for

$$\frac{dy}{dx} = \frac{-y+xy}{x-xy}$$

Using the chain rule, our system

$$\frac{dy}{dt} = y - yx \text{ and } \frac{dx}{dt} = -x + yx$$

gives that equation.

A. EULER'S METHOD ON A SYSTEM OF EQUATIONS

In this demonstration we compare the accuracy and speed of Euler's method to the Runge-Kutta method as we look at a second degree differential equation.

A second-order equation revisited

We have already solved the equation

$$\frac{d^2 y}{dt^2} + 2\frac{dy}{dt} + 2y = 0$$

in Chapter 24, writing it as two first-order equations. We will use both numeric methods on the new initial conditions $y(0) = 2$ and $y'(0) = 0$. In terms of this equation modeling a spring, this means that it has been dragged two units away from equilibrium and released with no initial velocity.

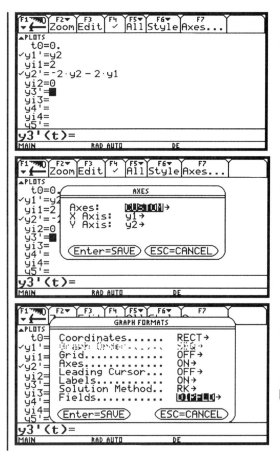

Change **MODE** to **DE** if needed and enter the second-order equation as a system of equations.

Use **F7** (**Axes...**) to change the axes so that **y1** is on the horizontal axis, **y2** on the vertical.

Check the **GRAPH FORMAT** settings by pressing ♦_F (from the Y= editor). The **Solution Method** should be **RK** and the **Fields** setting should be **DIRFLD**.

A. EULER'S METHOD ON A SYSTEM OF EQUATIONS

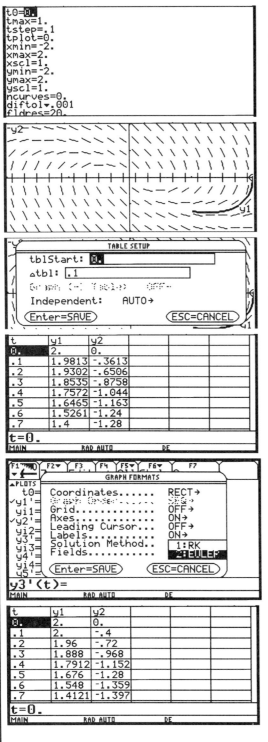

Now set the window as shown. We will only take 10 steps for our demonstration

Graphing shows the direction field and one phase plot.

To look at the values of y1 and y2 used in the graph, enter the table parameters shown and press ♦_TABLE.

You will see there is some delay as the values are calculated; it takes over 15 seconds.

Now we want to compare this to Euler's method. Go to the Y= editor screen and then press ♦_F. Select EULER and press ENTER twice to set the field and save the format settings.

Press ♦_TABLE again to see a new table. You can see there is little delay as the values are calculated; it takes under 5 seconds. This speed, though, comes at the price of some accuracy. You can see the difference in table entries, and the RK values are better.

Figure A Comparing of the accuracy and speed of Euler's method against the Runge-Kutta method for a second degree differential equation.

APPENDIX

Complex number form

Polar coordinates in the complex plane

Parametric graphing

Internet address information

Linking calculators

Linking to a computer

Troubleshooting

APPENDIX

Complex number form

The TI-92 accommodates complex numbers in the form $a + bi$ where $i = \sqrt{-1}$. Be sure to use the special complex i (2nd_i) and not the variable i. The need for parentheses becomes acute when writing complex number operations. As an example, the top two entries in the history area of Figure App.1 are the same sequence of numbers, but only in the second entry is the calculation clear. The next calculations in Figure App.1 show that some functions will report complex results when complex numbers are used as input. The ln first reports an error, but the same input in complex form produces an answer. Recall that solve will not accept complex numbers and finds only real solutions.

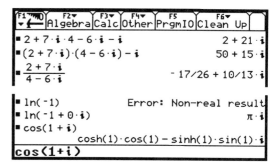

Figure App.1 Complex number arithmetic examples.

The 2nd_MATH 5:Complex menu has a list of operations that can be used on complex numbers. This menu and a calculation are shown in Figure App.2. The entry ln(-1) would normally give an error, but inside a complex number command it works fine.

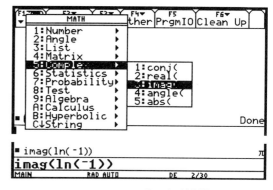

Figure App.2 Using the 2nd_MATH menu.

Polar coordinates in the complex plane

In MODE, the default Complex Format setting is Real, although Polar is an option. In most cases it is better to use the Real setting and convert numbers that need to be in polar format. Polar graphing is explained fully in Chapter 12 of the TI-92 *Guidebook*, so we give a very brief presentation here.

Coordinate conversion

Each point in the Cartesian rectangular coordinate system has a polar coordinate form (r,θ) where r is the distance to the origin and θ is a measure of rotation from the x-axis. The TI-92 representation of polar form is $re^{i\theta}$.

In Figure App.3 we see the coordinate conversion tools ▸`Polar` which can be found in the `2nd_CATALOG` under P. The ▸`Polar` command will put a rectangular coordinate complex number in polar form. With `Complex Format` set to `Real`, any polar entry will be converted to $a + bi$ format.

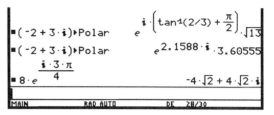

Figure App.3 Conversion techniques for complex numbers.

▶ *Tip: You will get an error message if you try to convert a real number to a complex one.*

▶ *Tip: The theta (θ) key is next to the M key.*

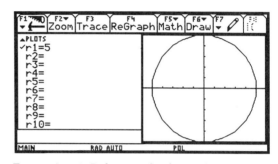

Figure App.4 Polar graph of a circle in a `ZoomDec` window.

Polar graphing

One of the important aspects of graphing in the polar or parametric form (discussed in the next section) is that the curve need not be a function; the graphs need not pass the vertical line test.

After changing the `MODE Graph` setting from `FUNCTION` to `POLAR`, the `Y=` screen shows `ri=` function definitions and `POL` is shown on the status line. In Figure App.4, a circle of radius 5 is drawn with a `ZoomDec` window.

In this graphing mode, the independent variable is θ. Figure App.5 shows the graph of $r = \theta$, first with the default coordinate setting

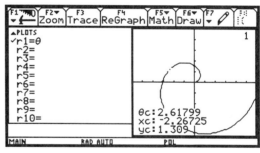

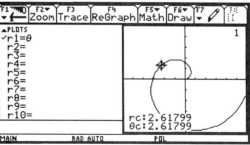

Figure App.5 The polar equation $r=\theta$ traced first with `Coordinates RECT`, then with `POLAR`.

(rectangular). Press ♦_F to access the GRAPH FORMATS screen and change the setting from RECT to POLAR; the second graph shows polar coordinates of the same graph. Notice in both settings that pressing the right arrow key while tracing increases θ-values which may correspond to moving left or right in *x*.

Parametric graphing

The second Graph option in MODE is PARAMETRIC. In this kind of graph, *x* and *y* are independently defined in terms of a variable *t* (usually thought of as time). The equation for a circle of radius 5 was easy in polar coordinates (Figure App.4). As a parametrically defined curve, a circle is composed of the sine and cosine functions. In Figure App.6 we use MODE to set PARAMETRIC (PAR is shown on the status line) and define the equation of a circle with radius 4. Notice the changes in the Y= editor: equations are now in pairs and the independent variable is t. We set a ZoomDec window with tstep=.1. Now F3 (TRACE) moves in steps of 0.1 for *t*, with $0 \leq t \leq 2\pi$.

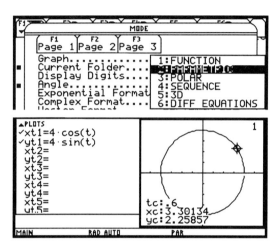

*Figure App.6 Select **PARAMETRIC**, define a pair of parametric equations, and graph in a standard window.*

Changing parameters

In Figure App.7 we look at the effect of changing the parameters of the equations defined in Figure App.6. The first curve appears the same. But by tracing you will see the effect of increasing the coefficient of t to 3: the curve is wrapped around the same path three times. You could also use TABLE to see this difference. In the other screens, a lower coefficient on the cosine creates an ellipse and a mixture of periods results in a beautiful pattern.

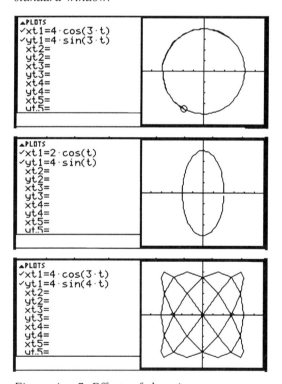

Figure App.7 Effects of changing parameters.

➤ *Tip: The only* Zoom *feature that will change the t-settings is* ZoomStd. *All others* Zoom *options will reset the window size, but not the t-values.*

Writing y = f (x) functions in parametric form

Any function $y = f(x)$ can be written in parametric form by setting xt1=t and yt1 to $f(t)$. We see in Figure App.8 that our first graph is restricted to the sine curve with $0 \leq t \leq 2\pi$ (the ZoomTrig setting). In the next graph, we changed t so that the graph extends across the window.

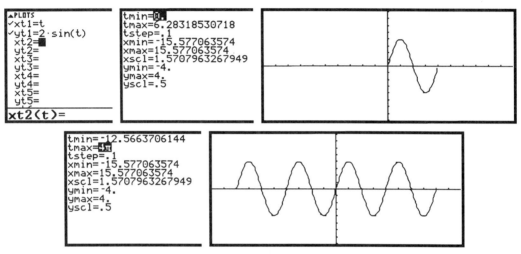

Figure App.8 Graphing y = f(x) parametrically.

Internet address information

The main internet address for Texas Instruments is

http://www.ti.com/

The calculator materials main menu can be obtained at

http://www.ti.com/calc/docs/

It is unlikely the above two entry addresses will change. Web page designs do change, but the following addresses are current as of April 1998 and will probably remain stable for finding the following topics.

Topic:	Address
Calculator Based Laboratory (CBL):	*http://www.ti.com/calc/docs/cbl.htm*
Calculator comparison:	*http://www.ti.com/calc/docs/gmtrx.htm*
Calculator support:	*http://www.ti.com/calc/docs/calcsupt.htm*

Classroom activities: *http://www.ti.com/calc/docs/activities.htm*
Frequently Asked Question(FAQs): *http://www.ti.com/calc/docs/faq.htm*
GraphLink: *http://www.ti.com/calc/docs/link.htm*
Guides to downloading: *http://www.ti.com/calc/docs/guides.htm*
Guidebook download: *http://www.ti.com/calc/pdf/gb/92gb.pdf*
New calculator news: *http://www.ti.com/calc/docs/calcnews.htm*
Program Archives: *http://www.ti.com/calc/docs/arch.htm*
Ranger (like the CBL): *http://www.ti.com/calc/docs/cbr.htm*
Resources: *http://www.ti.com/calc/docs/resource.htm*

There are discussion groups available; you will find information about these from the main screen of *http://www.ti.com/calc/docs/*.

Linking calculators

The essentials of linking are presented in Chapter 18 of the TI-92 *Guidebook* and will not be repeated here. But we include the following tips.

➤ *The end-jack must be pushed firmly into the socket. The calculator will turn on as it makes the proper connection.*

➤ *If you are experiencing difficulty connecting, turn off both calculators, check the connection, and then turn them on and try again. If available, try other cables or calculators.*

➤ *When selecting items, the check mark (✓) means selected, just as in the Y= editor.*

➤ *If you are required to drain your calculator memory before an exam, use F5 (All) 1:Select All to select all everything. Keep a copy on some other calculator. Even better is to store it on your computer, which is the next topic.*

Linking to a computer

The TI-Graph Link™ is a cable and software that connects a PC or Macintosh computer and your TI calculator. The software is available on the internet, so you can order just the cable. There are many advantages to using TI-Graph Link™:

- This is the best way to back up your work.
- It is the preferred way to write and edit programs.
- You can download and transfer programs from the internet archives.
- It allows you to capture the screen in a form for direct printing or use in a word processor.

Troubleshooting

Nothing shows on the screen
- Check the contrast.
- Check ON/OFF button.

The screen is frozen
- If there is a busy indicator on the status line, then the TI may be still calculating. Press ON if you can't wait.
- If there is a pause indicator on the status line, then the TI is paused from a program. Press ENTER to continue.
- Press ESC several times to exit any menus and then press ◆_HOME to return to the HOME screen.
- Go to the next section as a last resort.

Last resorts when stuck
Warning: This will erase all memory, including programs.
- Press and hold the 2nd and blue hand-key, then press and release the ON key.
- Pull out one of the AA batteries. Press and hold the (-) and) keys as you reinstall the battery. Continue to hold (-) and) for five more seconds, then release.

Y= Editor problems
- To edit a previous definition, you must first press either ENTER or F3 (Edit). Otherwise your keystrokes will be ignored.
- Be sure to use the correct independent variable for the graph mode you are using (FUNC uses x, PARAMETRIC uses t, etc.).

Nothing shows up on the graph screen
- Press TRACE to see if a function is defined but has values outside the window.
- There may be no functions selected. This may be caused by using NewProb.
- The function may be graphed right along either axis; reset the window.
- If there is a busy indicator on the status line, then the TI may be still calculating. Press ON if you can't wait.
- If there is a pause indicator on the status line, then the TI is paused from a program. Press ENTER to continue.

Nothing shows up in the table

- You may have the Ask mode set and need to either enter *x*-values or change to Auto in the TBLSET.
- Check to see that a function is selected. (NewProb deselects them all.)

Changes don't take

- Options, modes and settings made within a dialog box must be confirmed for each item and then again for the whole box. Press **ENTER** twice after making changes.
- To edit a previous definition in the Y= editor, you must first press either **ENTER** or **F3** (Edit). Otherwise your keystrokes will be ignored.
- Data cells in a table that have a formula in the header cannot be edited. Clear the formula.

I get an error message on the screen

This can cover the widest array of problems. Read the message carefully, and note where the cursor is positioned on the entry line. Press **ESC** and make a correction. If you have no idea what could have caused it, consult Appendix B of the TI-92 *Guidebook* for explanations of the error messages. The following are quite common:

- Parenthesis mismatch. Count and match parentheses carefully.
- Pasting a command in the wrong place. For example, a program name must be on a fresh line entry line.
- Circular definition error on the TI-92 without Plus. See page 19 of this book.

I'm getting a result but it is wrong

In all cases, check the pretty print version in the history area first.

- Parenthesis mismatch. Count and match parentheses carefully.
- Subtraction vs. negative symbol. For example, the subtraction sign cannot be used to enter a -10.
- Order of operations. For example, 1/2x is x/2 on a TI-92 and not the same as 1/(2·x).
- There may be an operation sign missing. For example, xy is a variable name, but x*y or x y is the product of x and y.
- Check that variable names you are expecting to be undefined have not been previously given a value. Use **F6** (Clean Up) to clear variables.
- Check any applicable default settings.

Split screen confusion

Use 2nd-QUIT twice to get back to a FULL screen. Press MODE and reset the options properly.

Common problems with solve

- The output may be correct but not in the expected algebraic form.
- The first entry must be an equation, not an expression, as is the case with zeros.
- The equation entry is incorrect. You should check in the history area to see that it is what you meant.
- A coefficient may have been previously defined.
- An expression or variable may be a reserved word. See page 491 in the TI-92 *Guidebook* for a list of these.

My program won't run

Program errors are difficult to diagnose. Open the program code with the sequence APPS 7 2. Add temporary displays and pauses to your code to check the progress and isolate the problem.

INDEX

Σ, 97, 143
$\int$, 101
$\approx$, 9
◆, 6
↑, 6
←, 7
2nd, 6
3D graphing, 52
@, 162
@n, 45
Anchorage annual rainfall, 127
Antiderivative, 105
Approximating area, 98
Approximations, 9
Arc length, 121
Asymptotic dangers, 32
AUTO Exact/Approx, 44, 87, 118
Average velocity, 56
Basic keys, 3
Boarding school epidemic, 185
Box with lid, 87
Bus stop problem, 90
Cabri Geometry, 90
CATALOG, 15
Chain Rule, 81
Circular definition, 19
CLEAR, 7
Cobb-Douglas, 52
comDenom, 128
Comparison test, 155
Complex numbers, 195
Complex numbers, 43
Compounding, 125
Concavity, 74
Conditional solutions, 45
Converging series, 141
Critically damped motion, 175
cSolve, 43

d limitations, 65
d and nDeriv, 66
d(∫(...)...) and ∫(d(...)...), 108
d(...), 64
Data editor, 177
Define, 17
Definite integral, 100
Degrees, 11
DEL, 7
Derivative, 61
　at a point, 62
　e^x, 71
　exact/numeric, 65
　graph of, 69
　matching graph, 69
　common functions, 72
　tangent function, 81
deSolve, 162
Differential vs. difference equation, 166
Differentiation rules, 78
Direction field, 187, 189
DIRFLD, 187
Discrete vs. continuous, 159
Drug dosage, 143
dy/dx, 63
Editing keys, 7
Entry line, 4
ENTER, 4
ENTRY, 8
ESC, 16
Euler's method, 166, 190
e^x series, 136
e^x, 108
e^x, derivative, 71
Exact derivative, 65
expand, 128
Extrema, 37
factor, 12

INDEX

..ults, 66
..g data, 180
..rce and pressure, 122
Format of numbers, 8
Formulas, 14
Fourier approximation function, 152
Function
    composite, 18
    editing inside a table, 23
    evaluating at a point, 18
    inverse, 39
    lists of values, 20
    new from old, 18
    piecewise, 151
    selecting and deselecting, 22
    table of values, 20
    TI-92 without Plus, 20
    user-defined, 148
    values on graph, 28
    zero from a table, 22
Fundamental Theorem, 105
Future value, 126
Geometry construction, 90
Graph, 24
    basics, 24
    behavior, 60
    derivative function, 69
    inaccurate, 34
    intersections, 37
    plots, 179
    split screen, 39
    text, 41
    3D, 52
Greatest equation ever written, 6
HOME screen, 4
Implicit differentiation, 94
Improper integrals 117
Inaccurate graph, 34
Indefinite integral, 105
Infinite number of solutions, 45
Infinite series, 155
Inflection, 89
INS, 7

Integrals
    applications, 121
    definite, 100
    indefinite, 105
    improper, 117
    infinite limit of integration, 117
    integrand infinite, 119
    partial fractions, 128
Internet addresses, 198
Intersection of graphs, 37
Inverse functions, 39
Ladder problem, 83
Learning curve, 159, 161
Left- and right-hand sums, 96, 98
Limit, one-sided, 61
Limit, 60
Limit, graphical/numerical, 57
Linear second-order equations, 173
Linking calculators, 199
Linking to a computer, 199
Local and long-term behavior, 60
Logistic curve, 74, 180
Logistic population, 177
Malthus, 18
Mean, 127
Menu use, 16
Minimum/Maximum, 37
MODE, 10
nDeriv, 66
nDeriv accuracy, 67
NewProb, 13, 45
nInt versus ∫, 109
Normal distributions, 126
Numeric derivative, 65, 68
Numeric solver, 47, 124
ON key, 3
One-sided limits, 61
Optimization, 83, 90
Overdamped motion, 174
Parametric graphing, 197
Partial fractions, 128
Phase plane, 184
Phase plots, 186, 189

Piecewise defined functions, 151
Piggy-bank vs. Trust, 146
Plot style, 34, 70
Plots of data lists, 179
Polar coordinates, 195
Polar graphing, 196
Predator-prey model, 188
Present value, 124
Product Rule, 78
Program entry error, 112
Programs, 110
Questionable accuracy, 104, 120
`QUIT`, 16
Quotient Rule, 79
Radians, 11
Regression line, 183
Riemann sums, 110
Roots, see zero
`rsum()` program, 112, 114
S-I-R model, 184
Savings accounts, 144
Scatterplot, 182
Scientific notation, 9
Screen contrast, 3
Second derivative table, 76
Second derivative, 73
Second derivative, concavity, 89
Second-order differential equations, 170
Selecting and deselecting a function, 22
Sequences, 144
Sequence graph, 161
Sequence table, 160
Series converging, 141
Simultaneous solutions to
    systems of equations, 46
Sine animation, 48
Sine series, 138
Slope fields, 164
Slope line, 63
`SLPFLD`, 163
`Solve` 41
Solving equations, 41

Solving non-polynomial equations, 44
Solving single variable equations, 41
Solving with complex solutions, 43
Speeding ticket, 59
Split screen graphing, 39
Square wave function, 150, 154
Standard deviation, 127
Statistical features, 128
Status line, 11
Stored values, 12
Sums, left- and right-hand, 96
Symbolic derivative at a point, 64
Symbolic derivative, 68
Systems of equations, 184
`Table`, 20
`Table: Independent: Auto Ask`, 21
Table function values, 20
Tangent derivative, 81
`TANLN`, 63
`Taylor`, 136
Taylor polynomials, 137
`Text editor`, 50
Text on graphs, 41
The alternating harmonic series, 140
TI-89, 3
TI-92 Plus, 3, 158
Time plots, 185, 188
`Trace`, 28
Triangle wave function, 151, 154
Troubleshooting, 200
Undefined results, 61
Under damped motion, 175
Units, 130-134
Unreliable results, 120
Unreliable table, 59
US population data, 177
User-defined functions, 148
Variables
    clearing, 14
    defined, 13
    symbolic, 12
    undefined, 13

INDEX

    24, 26
  ̣e, 30
  good, 32
  panning, 30
  settings, 25
xmax, 25
xmin, 25
xres, 25
xscl, 25
Y= editor, 14, 24
ymax, 25
ymin, 25
yscl, 25
Zero, 35
Zero from a table, 22
Zeros, 41, 89
Zoom options, 26, 35